Undergraduate Texts in Mathematics

Undergraduate Texts in Mathematics

Undergraduate Texts in Mathematics are generally aimed at third- and fourth-year undergraduate mathematics students at North American universities. These texts strive to provide students and teachers with new perspectives and novel approaches. The books include motivation that guides the reader to an appreciation of interrelations among different aspects of the subject. They feature examples that illustrate key concepts as well as exercises that strengthen understanding.

For further volumes:
http://www.springer.com/series/666

Steven Roman

Introduction to the
Mathematics of Finance

Arbitrage and Option Pricing

Second Edition

Springer

Steven Roman
Irvine, CA
USA

ISSN 0172-6056
ISBN 978-1-4899-8599-6   ISBN 978-1-4614-3582-2 (eBook)
DOI 10.1007/978-1-4614-3582-2
Springer New York Heidelberg Dordrecht London

Mathematics Subject Classification (2010): 91-01, 91B25

Printed on acid-free paper

Springer is part of Springer Science+Business Media (www.springer.com)

To Donna

Preface

This book has one specific goal in mind, namely to determine a *fair price* for a financial *derivative*, such as a stock option. The problem can be put in a very simple context as follows. Imagine that you are an investor in precious metals, such as gold or silver. Consider a one-ounce nugget of gold whose current value is $1800. The owner of this gold is willing to enter into a contract with you that gives you the right to buy the gold from him for $1750 at any time during the next month.

Obviously, the owner is not going to enter into such a contract for free, since he would lose $50 if you were to exercise your right immediately. But the owner will probably want more than $50, since there is a definite possibility that the price of gold will exceed $1800 over the next month.

On the other hand, there are limits to what you should be willing to pay for the *right to buy* the gold nugget. For instance, you would probably not pay $250 for this right. Assuming that both parties are eager to speculate (that is, gamble) on the future price of gold, there may be a price that both you and the owner of the gold will accept in order to enter into this contract. The purpose of this book is to build mathematical models that determine a *fair price* for such a contract.

In technical terms, the contract to buy the gold is a *call option* on gold, the buying price $1750 is the *strike price* and the date one month from today is the *expiration date* of the call option. Since the value of the contract at any given moment depends solely on the value of gold, the option is called a *derivative* and the gold is the *underlying asset* for the derivative. Our goal is to determine a fair price for this and other derivative financial instruments.

The intended audience of the book is upper division undergraduate or beginning graduate students in mathematics, finance or economics. Accordingly, no measure theory is used in this book.

It is my hope that this book will be read by people with rather diverse backgrounds, some mathematical and some financial. Students of mathematics may be well prepared in the ways of mathematical thinking but not so well prepared when it comes to matters related to finance (portfolios, stock options, forward contracts and so on). For these readers, I have included the necessary background in financial matters.

On the other hand, students of finance and economics may be well versed in financial topics but not as mathematically minded as students of mathematics. Nevertheless, since the subject of this book is the *mathematics* of finance, I have not watered down the mathematics in any way (appropriate to the level of the book, of course). That is, I have endeavored to be mathematically rigorous *at the appropriate level*. However, for the benefit of those with less mathematical background, I have made the book as mathematically self-contained as possible. Probability theory is ever present in the area of mathematical finance and in this respect the book is completely self-contained.

The Second Edition

This second edition is a complete rewriting of the first edition and has been influenced greatly by my having taught a class based on the first edition for the last five years running. In particular, the topic organization has been changed significantly, making the book flow much more smoothly. Most proofs have been rewritten and many have been improved significantly. The material on probability has been condensed into fewer chapters. The discussion of options has been expanded, including some information about the history of options and the reason why option pricing has become so important.

The discussion of pricing nonattainable alternatives has been expanded significantly. In particular, a new appendix has been added that contains proofs that the minimum dominating price of any nonattainable alternative is actually achieved by some dominating attainable alternative; that the maximum extension price is achieved by some nonnegative extension and that the minimum dominating price is equal to the maximum extension price. Finally, the material on the capital asset pricing model has been removed.

Organization of the Book

The book is organized as follows. The first chapter is devoted to the basics of stock options. In Chapter 2, we illustrate the technique of derivative asset pricing through the assumption of no arbitrage by pricing plain-vanilla forward contracts and discussing some simple issues related to option pricing, such as the put-call option parity formula.

Chapters 3 and 4 provide a thorough introduction to the topics of discrete probability that are needed for the subject at hand. Chapter 3 is an elementary and quite standard introduction to discrete probability and will probably be familiar to those who have had a course in basic probability. On the other hand, Chapter 4 covers topics that are generally not covered in basic probability classes, such as information structures, state trees, stochastic processes and martingales. This material is discussed only for discrete sample spaces and always keeping in mind that it is probably being seen by the reader for the first time.

Chapter 5 is devoted to the theory of discrete-time pricing models, where we discuss portfolios, arbitrage trading strategies, martingale measures and the first and second fundamental theorems of asset pricing. This prepares the way for the discussion in Chapter 6 on the binomial pricing model. This chapter introduces the important topics of drift, volatility and random walks.

In Chapter 7, we discuss the problem of pricing nonattainable alternatives in an incomplete discrete model. This chapter may be omitted if desired. Chapter 8 is devoted to optimal stopping times and American options. This chapter is perhaps a bit more mathematically challenging than the previous chapters and may also be omitted if desired.

Chapter 9 introduces the very basics of continuous probability. We need the notions of convergence in distribution and the Central Limit Theorem so that we can take the limit of the binomial model as the length of the time periods goes to 0. We perform this limiting process in Chapter 10 to get the famous Black–Scholes option pricing formula.

In Appendix A, we give optional background information on convexity that is used in Chapter 6. As mentioned earlier, Appendix B supplies some proofs related to pricing nonattainable alternatives.

A Word on Definitions

Unlike many areas of mathematics, the subject of this book, namely, the mathematics of finance, does not have an extensive literature at the undergraduate level. Put more simply, there are very few undergraduate textbooks on the mathematics of finance.

Accordingly, there has not been a lot of precedent with respect to setting down the basic theory at the undergraduate level, where pedagogy and use of intuition are (or should be) at a premium. One area in which this seems to manifest itself is the lack of terminology to cover certain situations.

Therefore, on rare occasions I have felt it necessary to invent new terminology to cover a specific concept. Let me assure the reader that I have not done this lightly. It is not my desire to invent terminology for any other reason than as an aid to pedagogy.

In any case, the reader will encounter a few definitions that I have labeled as *nonstandard*. This label is intended to convey the fact that the definition is not likely to be found in other books nor can it be used without qualification in discussions of the subject matter outside the purview of this book.

Thanks Be To ...

Finally, I would like to thank my students Lemee Nakamura, Tristan Egualada and Christopher Lin for their patience during my preliminary lectures and for their helpful comments about the manuscript of the first edition. Any errors in the book, which are hopefully minimal, are my responsibility, of course. The reader is welcome to visit my web site at *www.romanpress.com* to learn more about my books or to leave a comment or suggestion.

Contents

Part 3—The Black–Scholes Option Pricing Formula

9   Continuous Probability

10   The Black–Scholes Option Pricing Formula

Appendix A: Convexity and the Separation Theorem

Appendix B: Closed, Convex Cones

Notation Key and Greek Alphabet

$\langle , \rangle$: inner product (dot product) on $\mathbb{R}^n$

$\mathbf{1}$: the unit vector $(1, \ldots, 1)$

1_A^S or 1_A: indicator function for $A \subseteq S$

$\mathcal{A} = \{a_1, \ldots, a_n\}$: assets

C: price of a call

$C(B_k)$: the child subtree number of state B_k

$\mathcal{D}_i(B_k) = \{$descendents of B_k at level L_i, where $i \geq k\}$

e_i: the ith standard unit vector

$\mathcal{E}_P(X)$: expected value of X with respect to probability P

$\Phi^{(k,k+m)}$: trading strategy that locks in gain in Φ from time t_k to time t_{k+m}

$\Phi^{(k)}$: trading strategy that locks in gain in Φ from time t_k to time t_{k+1}

$\Phi[a_j]$: single-asset trading strategy

$\Phi[a_j, t_k, B]$: single-asset, single-period, single-state trading strategy

$H(B_k)$: the path number of state B_k

I_Y: Inner product by Y, that is, $I_Y(X) = \langle X, Y \rangle$

K (strike price)

μ_X: expected value of X

$\Omega = \{\omega_1, \ldots, \omega_m\}$: states of the economy

P: price of a put

$\mathcal{P}_i = \{B_{i,1}, \ldots, B_{i,m_i}\}$: state partition

$\mathbb{P}$: probability measure

Part(X): the set of all partitions of X

r: risk-free interest rate

RV(Ω): vector space of all random variables from Ω to $\mathbb{R}$

RV$^n(\Omega)$: vector space of all random vectors from Ω to $\mathbb{R}^n$

$\rho_{X,Y}$: correlation coefficient of X and Y

S: (price of stock or other asset)

$\sigma = (s_1, \ldots, s_m)$: state vector

σ_X^2: variance of X

$\sigma_{X,Y}$: covariance of X and Y

Θ_i: portfolio

$\mathcal{V}_0$: initial cost function

$\mathcal{V}_T$: payoff function

$\mathbb{X}_\tau$: the final payoff under a stopping time

Greek Alphabet

A α alpha	H η eta	N ν nu	T τ tau
B β beta	Θ θ theta	Ξ ξ xi	Υ υ upsilon
Γ γ gamma	I ι iota	O o omicron	Φ ϕ phi
Δ δ delta	K κ kappa	Π π pi	X χ chi
E ϵ epsilon	Λ λ lambda	P ρ rho	Ψ ψ psi
Z ζ zeta	M μ mu	Σ σ sigma	Ω ω omega

Introduction

Motivation

The subject of this book is *not* how to determine the value of a financial asset, such as a share of stock or a bar of gold, sometime in the future. Estimates of future value for such financial instruments are generally made using tools such as *fundamental analysis* (examining a company's balance sheet, income statements and cash flows), or *technical analysis* (drawing future conclusions from the price history of the asset) or some other mainly nonmathematical analysis.

Our goal in this book is to estimate the *current fair value* of the *option* to buy (or the option to sell) a given asset over some period of time in the future. This is done by assuming that the asset in question will have one of several possible values in the future and trying to determine a current fair value of the option based on these possible future values.

The option to buy (or the option to sell) a stock for a fixed value in the future is called a **stock option**. An option to buy is called a **call** and an option to sell is called a **put**. The buying (or selling) price is called the **strike price**. As we will see, options can be based on assets other than stocks, although stock options are by far the most common form of option.

If a call has a strike price that is less than the current market value of the asset, then the option has immediate value and is said to be **in the money**. Similarly, a put is in the money at a given time if the strike price is greater than the current market price of the asset.

Since the invention of stock options in the 1920s, the granting of these financial instruments has played a very large role an as incentive for hiring and retaining company executives. This is because for several decades the granting (gifting) of stock options (in the form of calls) has had a significant tax advantage over direct cash compensation. In fact, by the 1950s, option grants accounted for almost one-third of all executive compensation in large companies.

Indeed, as late as the 1990s, the federal government encouraged the use of stock options as a form of executive compensation, as illustrated by the following facts:

1) In 1993, in an effort to limit executive pay, the IRS prohibited companies from deducting more than 1 million dollars in annual compensation for company executives.
2) In 1994, Congress defeated a proposal by the Securities and Exchange Commission that would have required companies to treat the granting of stock options as an expense and deduct it from the company's earnings.
3) The tax law allowed a tax deduction whenever stock options were exercised under which the *company* could deduct from its income an amount equal to the amount of an *employee*'s gain from option compensation.

However, in the atmosphere of these rather permissive rules, some companies began to invent creative ways to manipulate the situation. Here are some examples.

1) *Backdating*: Stock options are granted based on a date prior to the time of granting, when the stock price was lower, making the options effectively in the money when they might not otherwise have been in the money. Several hundred companies appear to have backdated stock options.
2) *Repricing*: The option's strike price is lowered *retroactively* if the option fails to be in the money during the exercise period. Studies indicate that approximately 11 percent of companies repriced options at least once between 1992 and 1997.
3) *Reloading*: Options that are exercised by the employee are automatically replaced by options at a lower strike price (but typically in fewer numbers). By 1999, nearly 20 percent of large companies offered reloading plans.

Starting in the 1990s, steps were taken by the federal govenment to address the issue of granting in-the-money options to avoid payment of taxes. These include the following:

1) The Financial Accounting Standards Board (FASB) Statement No. 123 (issued October 1995) requires that a company's financial statements include certain disclosures about stock-based employee compensation. In particular, *granted stock options must be assigned a fair value using some pricing model and booked as an expense by the company*.
2) The Sarbanes–Oxley Act of 2002 prohibits the backdating of options and strengthens the requirements for reporting stock option grants for public companies.
3) The IRS changed the tax laws with regard to the granting of in-the-money stock options.

The requirements contained in the FASB statement brought to the forefront the problem that is the subject of this book: namely, the problem of assigning a fair value to (stock) options.

One might at first think that the issue is simple: just set the fair value of an option to its current market value. However, the problem is that in general, the options granted as employee compensation do not exist on the open market and therefore do not have a market value! Thus, we must turn to mathematical models for the purpose of assessing fair value.

With this motivation in mind, let us take a fresh look at the problem.

The Derivative Pricing Problem

A **financial security** or **financial instrument** is a legal contract that conveys *ownership* (as in the case of a stock), *credit* (as in the case of a bond) or *rights to ownership* (as in the case of a stock option). When a financial security is traded, the buyer is said to take a **long position** in the security and the seller is said to take the **short position** in the security. The two positions are said to be **opposite positions** of one another.

Some financial securities have the property that their value depends upon the value of another security. In this case, the former security is called a **derivative** of the latter security, which is then called the **underlying security** or just the **underlying** for the derivative. The most well-known examples of derivatives are ordinary stock options (puts and calls). In this case, the underlying security is a stock.

However, derivatives have become so popular that they now exist based on more exotic underlying financial entities, such as interest rates and currency exchange rates. It is also possible to base derivatives on other derivatives. For example, one can trade options on futures contracts. Thus, a given financial entity can be a derivative under some circumstances and an underlying under other circumstances.

In fact, one can create a financial derivative based on any quantity $Q(t)$ that varies in a random (nondeterministic) way with time t. To illustrate, let t_0 be the current time and let $t_1 > t_0$ be a time in the future. Consider a financial instrument whose terms as as follows. At time t_1, if the change in value

$$A = Q(t_1) - Q(t_0)$$

is positive, then the seller pays the buyer the amount A. If not, then the seller pays nothing to the buyer.

This is a financial derivative since its value at time t_1 depends on the value Q of the underlying. Moreover, since there is *risk* involved in selling such an instrument, the seller will not be willing to enter into such a contract without some monetary compensation at the time t_0 of formation of the contract. Moreover, the buyer should be willing to pay something to the seller in order to acquire the possibility of receiving a payoff $A > 0$ at time t_1. The question is: "What is a fair price for this derivative?"

Determining a fair value for a derivative is called the **derivative pricing problem** and is the central theme of this book.

As a more concrete example, suppose that IBM is selling for $100 per share at this moment. A 3 month *call option* on IBM with *strike price* $102 is a contract between the buyer and the seller of the option that says that the buyer may (but is not required to) purchase 100 shares of IBM from the seller for $102 per share *at any time* during the next 3 months.

Of course, at this time, the buyer will not want to *exercise* the option, since he presumably has no desire to buy the stock for $102 per share from the seller when he can buy it on the open market for $100 per share. But if the price of IBM rises above $102 during the 3 month period, the buyer may very well want to exercise the call and buy the stock at $102 per share. Thus, the call option has some value and so the seller will want some monetary compensation to enter into this contract with the buyer. The question is: "How much compensation?"

The only time at which the derivative pricing problem is easy to solve is at the *time of expiration* of the derivative. In the previous example, if at the end of the 3 month period, IBM is selling for $103, then the value of the call option at that time is $103 − $102 = $1 (ignoring additional costs, such as transaction costs and commissions). However, at any earlier time, there is uncertainty about the future value of the stock price and so there is uncertainty about the value of the option.

Assumptions

Financial markets are complex. As with most complex systems, creating a mathematical model of a financial system requires making some simplifying assumptions. In the course of our analysis, we will make several such assumptions. For example, we will assume a **perfect market**; that is, a market in which

- there are no commissions or transaction costs,
- the lending rate is equal to the borrowing rate,
- there are no restrictions on short selling (defined later in the book).

Of course, there is no such thing as a perfect market in the real world, but this assumption will make the analysis considerably simpler and will also let us concentrate on certain key issues in derivative pricing.

In addition to the assumption of a perfect market, we also assume that the market is **infinitely divisible**, which means that we can speak of, for example, $\sqrt{2}$ or $-\pi$ shares of a stock. We will also assume that the market is **frictionless**; that is, all transactions take place immediately, without any external delays.

Risk-free Asset

We will also assume that there is always available a **risk-free asset**; that is, a particular asset that cannot decrease in value and generally increases in value. Furthermore, the amount of the increase over any given time interval is *known in advance*. Practical examples of securities that are generally considered risk-free assets are U.S. Treasury bonds and federally insured bank deposits.

For reasons that will become apparent as we begin to explore financial models, it is important to keep separate the notions of the *price* of an asset and the *quantity* of an asset and to assume that it is the *price* of an asset that changes with time, whereas the quantity only changes when we deliberately change it by buying or selling the asset.

Accordingly, one simple way to model the risk-free asset is to imagine a special asset with the following behavior. At the initial time t_0 of the model, the asset's price is 1. During a given time interval $[t_1, t_2]$, the asset's price increases by a factor of $e^{r_1(t_2 - t_1)}$, where r_1 is the **risk-free rate** for that interval.

It is traditional in books on the subject to model the risk-free asset as either a bank account or a risk-free bond. For a normal bank account, however, there is an issue that must be considered: namely, it is not the value of the units (say dollars) that change but the quantity. For example, if we deposit \$10 (10 units of dollar) in an account at time t_0 then after a period of 5% growth we have 10.5 units of dollar, not 10 units of dollar each worth 1.05. This issue must be kept in mind when using a bank account rather than a bond.

We will assume throughout the book that it is possible to buy or sell any amount of the risk-free asset.

Arbitrage

The term *arbitrage* suffers from a bit of a dichotomy. In a general, nontechnical sense, the term is often used to signify a condition under which an investor is *guaranteed* to make a profit regardless of circumstances.

The more commonly adopted technical use of the term is a bit different. An **arbitrage opportunity** is an investment opportunity that is guaranteed not to result in a loss and *may* (with positive probability) result in a gain. Note that the gain is not guaranteed, only the lack of loss is guaranteed. For example, a game in which we flip a fair coin once and get 1 dollar if the result is heads but nothing if the result is tails might not be considered arbitrage in the nontechnical sense but is definitely arbitrage in the technical sense. After all, who would not enter into such a game for free? Actually, one should be willing to play this game for any initial fee less than 50 cents, since the expected return will be positive. However, if there is any fee involved, the game is no longer an arbitrage opportunity, since a loss is now possible.

It is important to note that we must be very careful how we measure gain when assessing arbitrage. For instance, if $100 today grows to $100.01 in a year, is this true gain? Put another way, would you make this investment? Probably not, because there are probably risk-free alternatives, such as depositing the money in a federally insured bank account that will produce a larger gain.

As we will see, the key principle behind derivative pricing (or indeed any asset pricing) is that *market prices will adjust in order to eliminate arbitrage*; that is, if an arbitrage opportunity exists, then prices will be adjusted to eliminate that opportunity.

As a simple example, suppose that gold is priced at $980.10 per ounce in New York and $980.20 in London. Then investors could buy gold in New York and sell it in London, making a profit of 10 cents per ounce (assuming that transaction costs do not absorb the profit). However, purchasing gold in New York will drive the New York price higher and selling gold in London will drive the London price lower. As a result, the arbitrage opportunity will disappear.

This leads us to the fundamental principle of asset pricing:

No-arbitrage Pricing Principle: *As a consequence of the tendency to an arbitrage-free market equilibrium, it only makes sense to price securities under the assumption that there is no arbitrage.* □

Implementing the no-arbitrage pricing principle for pricing is actually quite easy in theory. Imagine two portfolios of financial assets. Let us refer to these portfolios as Portfolio A and Portfolio B. Let us also consider two time periods: the initial time $t = 0$ and a final time $t = T > 0$.

Each portfolio has an initial value and a final value or *payoff*. Let us denote the initial value of the two portfolios by $\mathcal{V}_{A,0}$ and $\mathcal{V}_{B,0}$ and the final values by $\mathcal{V}_{A,T}$ and $\mathcal{V}_{B,T}$. The values of Portfolio A are shown in Figure 1. A similar figure holds for Portfolio B.

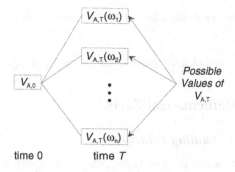

time 0   time T

Figure 1: The values of Portfolio A

As can be seen in the figure, Portfolio A has a *known* initial value $V_{A,0}$. On the other hand, the final value of Portfolio A is unknown at time $t = 0$. In fact, we assume that this value depends on the state of the economy at time T, which can take one of n possible values $\omega_1, \ldots, \omega_n$. Thus, the final value $V_{A,T}$ is actually a *function* of these states. Similarly, we assume that the initial value of Portfolio B is known and that the final value is a function of the possible states of the economy.

Now, consider what happens if Portfolios A and B have exactly the same payoffs *regardless of the state of the economy;* that is, if

$$V_{A,T}(\omega_i) = V_{B,T}(\omega_i)$$

for all $i = 1, \ldots, n$. The no-arbitrage pricing principle then implies that the initial values must be equal, that is

$$V_{A,0} = V_{B,0}$$

For suppose that $V_{A,0} > V_{B,0}$. Then under the assumption of a perfect market, an investor can purchase the cheaper Portfolio B and sell the more expensive Portfolio A, pocketing the positive difference $V_{A,0} - V_{B,0}$. At time T, *no matter what state the economy is in*, the investor receives the common final value of the portfolios and must pay out the same amount. Thus, he loses nothing at the end and can keep the initial profit. This is arbitrage in the strongest sense, namely, a *guaranteed* profit.

This approach can be used to determine an initial value of an asset, such as a derivative, whose final payoff is known. To price the asset, all we need to do is find a portfolio that has the same final payoff function as the asset we wish to price, but has a known initial value. This is called a *replicating portfolio*. It follows that the initial value of the asset in question must be equal to the initial value of the replicating portfolio.

The no-arbitrage pricing principle can be used in other ways to determine prices. For example, if the initial values of two portfolios are equal, then it cannot be that one portfolio *always* yields a higher payoff than the other, regardless of the state of the economy.

We will see many examples of the use of the no-arbitrage pricing principle throughout the book.

Miscellaneous Mathematical Facts

The Fundamental Counting Principle

Let $T_1, T_2, \ldots, T_k$ be a sequence of tasks with the property that the number of ways to perform any task in the sequence does not depend on how the previous tasks in the sequence were performed. Then, if there are n_i ways to perform the ith task T_i, for all $i = 1, 2, \ldots, k$ the number of ways to perform the entire sequence of tasks is the product $n_1 n_2 \cdots n_k$. For instance, if you are considering buying one of five different stocks and one of six different bonds, then there are $5 \cdot 6 = 30$ ways to buy one stock and one bond.

Permutations

Let S be a set of size n. An ordered arrangement of the elements of S is called a **permutation** of S. The **size** of each permutation is also n. For example, there are 6 permutations of the set $S = \{a, b, c\}$:

$$abc, acb, bac, bca, cab, cba$$

More generally, an ordered arrangement of size $k \le n$ of elements of S is called a **permutation** of size k taken from S. For instance, if $S = \{a, b, c, d\}$, then

$$adb \quad \text{and} \quad cda$$

are permutations of size 3. The number of permutations of a set is easily determined using the fundamental counting principle.

Theorem 1
1) *The number of permutations of size n is*

$$n! = n(n-1)\cdots 2 \cdot 1$$

*The number $n!$ is called n **factorial**. For consistency, we set $0! = 1$.*
2) *More generally, the number of permutations of size k, taken from a set of size n is*

$$n(n-1)\cdots(n-k+1) = \frac{n!}{(n-k)!}$$

Proof. Part 1) is a special case of part 2), since taking $k = n$ in part 2) gives $n!$. As to part 2), there are $n = n - 0$ ways to choose the first object in the permutation. Then there are $n - 1$ choices for the second object, $n - 2$ choices for the third object and so on. For the last object, there are $n - k + 1$ choices. The fundamental counting principle then gives the result.□

Combinations

Unordered arrangements of objects are better known as *subsets*. They are also called combinations. Specifically, an unordered arrangement of size k, taken from a set of size n, is called a **combination** of size k. In order to describe the number of combinations (subsets) of a set, we need the following concept.

Definition *The expression*

$$\binom{n}{k} = \frac{n!}{k!(n-k)!}$$

is called a **binomial coefficient**. The binomial coefficient $\binom{n}{k}$ is also denoted by $C(n, k)$.□

Theorem 2
1) The number of subsets of size k, taken from a set of n objects is $C(n, k)$.
2) The number of subsets of all sizes of a set of size n is 2^n.
Proof. Each combination of size k leads to $k!$ permutations when we order the objects in the combination. Hence, the number of permutations of size k is equal to $C(n, k) \cdot k!$, This gives the equation

$$C(n, k) \cdot k! = \frac{n!}{(n-k)!}$$

and solving for $C(n, k)$ proves part 1). As for part 2), we can use the fundamental counting principle. Let S be a set of size n. Arrange the elements of the set S in a row, as in

$$e_1 e_2 \cdots e_n$$

We can form a subset of S by deciding whether or not to include e_i in the subset for each element e_i. This requires making n choices, each of which has two possibilities: include or exclude. Hence, there are 2^n ways to make these choices, that is, there are 2^n subsets of S.□

Note that

$$\binom{n}{2} = \frac{n(n-1)}{2}$$

and that

$$\binom{n}{k} = \binom{n}{n-k}$$

The binomial coefficients play a role in the binomial formula

$$(a+b)^n = \sum_{k=0}^{n} \binom{n}{k} a^k b^{n-k}$$

Miscellanea

We will let $\mathbb{R}$ denote the set of real numbers. It will be convenient throughout the book to use the following notation:

$$X^+ = \max\{X, 0\}$$

where X is any number or function.

The **inner product** or **dot product** of two vectors $X = (x_1, \ldots, x_n)$ and $Y = (y_1, \ldots, y_n)$ in $\mathbb{R}^n$ is defined by

$$\langle X, Y \rangle = x_1 y_1 + \cdots + x_n y_n = \sum_{i=1}^{n} x_i y_i$$

Part 1—Options and Arbitrage

Chapter 1
Background on Options

We begin with a discussion of the basic properties of stock options. Readers who are familiar with these types of derivatives will want merely to skim through the chapter to synchronize terminology.

Stock Options

Stock options take two forms: *put options* (*puts*) and *call options* (*calls*). Here are the definitions.

Definition *A* **stock option** *is a contract between the* **writer** (**seller**) *of the option and the* **buyer** *of the option. The writer has a* **short position** *and the buyer has a* **long position**. *Every option has an* **underlying stock**, *an* **expiration date** *and a* **strike price**, *also called a* **striking price** *or* **exercise price**.
1) *In a* **call option**, *the buyer has the right to buy the underlying stock from the writer at the strike price K per share.*
   a) *In a* **European call**, *the right to buy can only be exercised on the expiration date of the call.*
   b) *In an* **American call**, *the right to buy can be exercised at any time on or before the expiration date of the call.*
2) *In a* **put option**, *the buyer has the right to sell the underlying stock to the writer at the strike price K per share.*
   a) *In a* **European put**, *the right to sell can only be exercised on the expiration date of the call.*
   b) *In an* **American put**, *the right to sell can be exercised at any time on or before the expiration date of the call.* □

We will generally reserve the letter K for the strike price of an option and the letter S for the price of the underlying stock. The cost of a call will be denoted by C and the cost of a put by P.

Although it will not be required for our mathematical analysis, we want to give some details about how stock options work.

Exchanges

Most but not all stocks have associated options. Options on major stocks are generally traded through an *options exchange*, the largest of which is the **Chicago Board of Options Exchange** (**CBOE**). The exchange determines the terms of an option, such as the expiration date and strike price. Generally speaking, options are traded in **round lots** of 100 underlying shares; that is, each options contract is a contract to buy or sell 100 shares of the underlying stock. (Contract sizes can vary when the underlying stock has undergone a stock split.) However, we will assume for our mathematical models that any *real number* of options can be purchased.

Option Terminology

The puts and calls form the two **classes** of options for a given underlying stock. An **option series** within a class is a particular strike price and expiration date. For example, one option series is

IBM JAN 50 CALLS

where 50 is the strike price (in dollars) and January is the expiration month.

Expiration Dates

The last trading day of an option is the third Friday of the expiration month and the option actually expires on the following Saturday. Every stock option is on one of three **expiration cycles**, which consists of one month per quarter, equally spaced 3 months apart, but starting at different months:

1) *January cycle*: Jan, Apr, July, Oct
2) *February cycle*: Feb, May, Aug, Nov
3) *March cycle*: Mar, June, Sept, Dec

If the expiration date for the current month has not passed, then there exist options that trade with expiration dates in the current month, the next month and the following two months of the cycle for that underlying. If the expiration date for the current month has passed, then there exist options that trade for the next month, the month after that and the following two months in the cycle.

For example, IBM is on the January cycle. At the beginning of January, there are options that expire in January, February, April and July. Late in January, there are options that expire in February, March, April and July. At the beginning of May, options expire in May, June, July and October.

Longer-term options are available on some stocks. These are called **LEAPS** (long-term equity anticipation securities). They have expirations up to 3 years in the future and expire in January.

Strike Prices

The CBOE normally sets the strike prices for its options so that they are spaced $2.50, $5 or $10 apart. Stocks at lower prices have smaller spaces between strike prices. When options with a new expiration date are introduced, the CBOE usually introduces two or three options with strikes nearest to the current stock price. If the price moves outside this range, new strikes may be introduced. For example, if new October options are offered on a stock currently priced at $84, then options striking at $80, $85 and $90 might be created. If the price rises above $90, a new strike at $95 might be introduced.

Option Symbols

Every stock has a symbol used for identification. For example, IBM is the symbol for International Business Machines and GE is the symbol for General Electric Corporation. Options also have symbols. Up until February 12, 2010, the symbols used for options were confusing, to say the least. Starting February 12, 2010 and fully implemented by May 2010, options symbols have been standardized to the following form:

1) The first portion of the symbol is the underlying company's root symbol.
2) This is followed by two characters each for the maturity year, month and day.
3) This is followed by a "C" for call or a "P" for put.
4) The final portion of the symbol is the strike price. Here five characters are devoted to the dollar portion and three characters to any decimal portion.

The only "catch" in constructing option symbols is that one needs to know the date of the Saturday following the third Friday of the expiration month.

For example, in 2010, an IBM July 125 call has expiration date July 17, 2010 and so the symbol is

IBM100717C00125000

It is probably worth noting that individual brokerage houses have created their own option symbols (unfortunately). For example, the Charles Schwab symbol for the option above is

IBM 07/17/2010 125.00 C

and Fidelity Investments recognizes the option in the form

-IBM100717C125

whereas Yahoo! Finance recognizes the standard symbology.

The Role of the Options Clearing Corporation

When an investor instructs his broker to buy or sell an option, the broker transmits this request to the firm's *floor broker* on the appropriate options exchange, who attempts to locate another floor broker (or other official) who has instructions to perform the opposite transaction on behalf of another investor. The trade is then made and both brokers record the details of the transaction. This entire process generally takes only a few minutes.

However, under this simple scenario, the buyer of the option would have to trust the seller to make good on his obligation to buy/sell the underlying. It is the role of the **Options Clearing Corporation** (**OCC**) to remove this dependency. At the end of the day, the OCC examines all of the day's trading, matching each sale with the corresponding purchase. It then inserts itself between the buyer and the seller, playing the role of the buyer for the seller and the role of the seller for the buyer. Hence, each investor deals only with the OCC (indirectly) and not the other investor. The OCC has sufficient resources to make good on any amounts owed as well as to enforce any collection, should that be required.

The OCC also plays a role in the exercise of an option. When an investor notifies a broker that he wants to exercise an option, the broker places the exercise order with the OCC. The OCC randomly selects a member brokerage firm that has at least one writer of that option. The member brokerage firm, using a predefined algorithm, selects a particular investor who has written the option. This investor is said to be **assigned**. Thus, an investor who has sold an American option never knows when he may be required to make good on his obligation.

Open Interest

An option has one of three fates: it *expires* without exercise, it is *exercised* or it is *closed* by an *offsetting transaction*. An investor who owns an option can close his position by issuing a special offsetting transaction with the same class, underlying, strike price and expiration but opposite position (long/short). The buyer closes an open option by selling, the writer closes by buying. Thus, every option transaction is one of the following four types:

1) buy to open
2) buy to close
3) sell to open
4) sell to close

The **open interest** is the number of outstanding contracts. When an option contract is traded, if both investors are opening, then the open interest increases by 1, since there is a new contract. If one trader is opening and one trader is closing, the existing contract is passed from the closer to the opener and so the open interest is unchanged. If both investors are closing, the open interest declines by 1.

For example, Figure 1.1 shows two investors. In the first figure, investor A has bought an option and investor B has sold an option. The open interest is 1, which is one-half the number of arrows, since each pair of arrows (pointing in opposite directions) corresponds to a single contract.

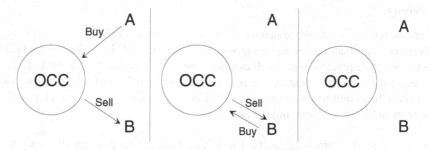

Figure 1.1

If A decides to sell to close, the OCC will find a matching buyer, say B. If B buys to open, then the end result is shown in the middle portion of Figure 1.1 and the open interest is still 1. If instead B buys to close, then the result is shown in the right-hand portion of Figure 1.1 and the open interest is now 0.

Underlyings

As mentioned earlier, options exist on many different types of underlying investments other than common stocks. For example, options exist on foreign currencies, futures contracts and stock indices.

The most popular index options are on the S&P 500, S&P 100, Nasdaq 100 and Dow Jones Industrial Index. Some options are European (e.g. S&P 500) and some are American (e.g. S&P 100). An index option grants the right to buy 100 times the index value for the strike price. Settlement is always in cash, not in stock. For example, if a call on the S&P 100 with strike price 980 is exercised when the index is 992, the writer must pay the buyer $12 \times 100 = 1200$ dollars.

The Purpose of Options

Options are primarily used for *hedging* and for *speculation*. A **hedge** is an investment that reduces the risk in an existing position. To illustrate the hedging feature of an option, suppose an investor currently owns 1000 shares of XYZ, whose current price is $88 per share. The investor suspects that there might be a significant drop in the stock price in the near future (perhaps some announcement is pending that could dramatically affect the stock price).

So to hedge against this possibility, the investor buys a three-month put with strike price $85, which gives him the right to sell the stock at $85 per share for the next 3 months. Thus, if the stock price drops below $85, the investor can exercise the option, thereby limiting his loss to $3 per share. The price paid for this hedge is the price of the put, which is currently selling for $1.50 per share.

Thus, a $1500 outlay will protect an $88,000 investment against more than a $3000 loss over the 3 month period.

Leverage

Options have one major advantage over owning the underlying asset, namely, **leverage**. To illustrate leverage, suppose that a stock ABC is selling for $90 per share. A small investor with $450 can purchase only 5 shares of the stock. If the investor feels that the stock price is about to rise significantly, then the use of options allows him to leverage his bankroll and speculate on the stock in a much more meaningful way than buying the shares.

For example, the current price of a 1 month call with strike price of $90 is $3.80. Thus, the investor can purchase 118 such calls (ignoring commissions). If the price of ABC is $95 at exercise time the profit on 5 shares would be only $25 whereas the profit on 118 calls would be $140. The return is thus over 31% on the investment in options, whereas it is less than 6% for the stock investment! This is leverage.

Of course, the downside to the call options is that if the stock does not rise before the expiration date, the investor will receive nothing from the options and will have lost the price of these options, whereas the stockholder still owns the stock.

Profit and Payoff Curves

Generally speaking, when the expiration date arrives, the owner of an option will exercise that option if and only if there is a positive return. Thus, if the strike price of the option is K and the spot price (current price) of the stock is S, the owner of a call will exercise the option if $K < S$ and the owner of a put will exercise the option if $K > S$. The following terms are used to describe the various possibilities.

Definition *A call option is*
1)   **in the money** *if* $K < S$
2)   **at the money** *if* $K = S$
3)   **out of the money** *if* $K > S$
A put option is
4)   **in the money** *if* $K > S$
5)   **at the money** *if* $K = S$
6)   **out of the money** *if* $K < S.\square$

It is important to note that just because an option is in the money does not mean that the owner makes a *profit*. The problem is that the initial cost (as well as any commissions, which we will ignore throughout this discussion) may outweigh the return gained from exercising the option. In that case, the investor will still execute because the positive return will help reduce the overall loss.

Figure 1.2 shows the *payoffs* (ignoring costs) for each option position. The horizontal axis is the stock price at exercise time and all line segments are either horizontal or have slope ±1.

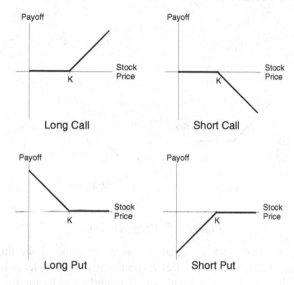

Figure 1.2: Payoff curves

The payoff formulas are actually quite simple. For a long call, if the stock price S satisfies $S \geq K$, then the payoff from exercising the call is $S - K$ whereas if $S < K$, then the call will expire worthless and so the payoff is 0. Thus, the payoff is

$$\text{Payoff(Long Call)} = (S - K)^+$$

where

$$X^+ = \max\{X, 0\}$$

for any number X. On the put side, we have

$$\text{Payoff(Long Put)} = (K - S)^+$$

Figure 1.3 shows the *profit* curves, which take into account the cost of option. (As mentioned, we will ignore all commissions.)

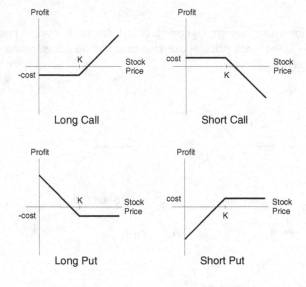

Figure 1.3: Profit curves

These payoff curves are very informative. Here are some of the things we can immediately see from these curves. Let K be the strike price, let S be the strock price, let C be the initial cost of a call per share and let P be the initial cost of a put per share.

Long Call

- Limited downside: The downside is limited to the cost C of the call.
- Unlimited upside: The upside is effectively unlimited, since there is no limit to the price of the stock.
- Optimistic (**bullish**) position: The buyer hopes the stock price will rise.
- Break-even point: The buyer breaks even if $S = K + C$. (Here we ignore the time value of money.)

Short Call

- Unlimited downside: The downside is effectively unlimited because there is no limit to the price of the stock.
- Limited upside: The upside is limited to the selling price C of the call.
- Pessimistic (**bearish**) position: The seller hopes the stock price will fall.
- Break-even point: The seller breaks even if $S = K + C$.

Long Put

- Limited downside: The downside is limited to the cost P of the put.
- Limited upside: The upside is also limited because the stock price can only fall to 0, in which case the profit is equal to $K - P$.
- Pessimistic (bearish) position: The buyer hopes the stock price will fall.

- Break-even point: The buyer breaks even if $S = K - P$.

Short Put

- Limited downside: The downside is limited because the stock price can only fall to 0, in which case the loss is equal to $K - P$.
- Limited upside: The upside is also limited to the selling price P of the put.
- Optimistic (bullish) position: The seller hopes the stock price will rise.
- Break-even point: The seller breaks even if $S = K - P$.

Setting aside for the moment the risk factor, we can also say the following:

- Even though a short call and a long put are both bearish positions, there is a difference. If we believe that a stock's price will settle near the strike K, then a short call is more advantageous that a long put, which will still result in a loss due to the cost of the put. However, if we believe that a stock's price will decline sharply, then a long put is more advantageous.
- Similarly, if we believe that a stock's price will settle near the strike K, then a short put is more advantageous than a long call.

Covered Calls

We have said that a short call position has an unlimited downside because the stock price can theoretically rise without bound and so if the seller needs to buy the shares at exercise time, he has a potentially unlimited risk.

One way to mitigate this risk is to buy the shares at or before the time that the option is sold. If the seller of a call option owns the stock, the call is said to be **covered**. Writing covered calls is far safer than writing **uncovered** (also called **naked**) calls. For this reason, a brokerage house places much stronger restrictions on allowing the sale of uncovered calls than on the sale of covered calls.

Similarly, selling an uncovered put has a potentially large downside, since the seller may be required to buy the stock for the strike price K, even if the stock price goes to 0. Accordingly, to **cover a put**, the writer sells the stock short (described in detail a bit later in this chapter). Then if the put is exercised, the writer can use the stock he is forced to purchase to unwind the short sale. In this way, the writer has protected himself up to the initial price S of the stock, which is received from the short sale. Thus, the downside is at most $K - Se^{rt}$, where t is the time to maturity of the option and r is the risk-free rate.

Profit Curves for Option Portfolios

An **option portfolio** consists of a collection of options of varying types. The following example shows how to obtain the profit curve for a simple option portfolio.

Example 1.1 Consider the purchase and sale of options, all with the same expiration date, given by the following expression:

$$- P_{100} + P_{120} + 2C_{150} - C_{180}$$

This position is: short a put with strike price 100, long a put with strike price 120, long two calls with strike price 150 and short a call with strike price 180. The overall payoff curve can be obtained from the individual payoff curves by plotting them all on a single set of coordinates, as shown in Figure 1.4. Note that it is simpler to ignore all costs in drawing the curves and then simply translate the final curve an amount equal to the total cost for all the options in the portfolio, which in this case is

$$-\text{Cost}(P_{100}) + \text{Cost}(P_{120}) + 2\text{Cost}(C_{150}) - \text{Cost}(C_{180}) \qquad \Box$$

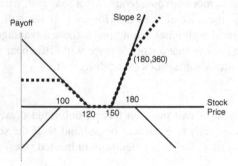

Figure 1.4: Payoff curve

The Time Value of an Option

The payoff $(S - K)^+$ of a call and the payoff $(K - S)^+$ of a put are also referred to as the **intrinsic value** of the option. However, the *market price*, also called the **premium** of an option, is seldom equal to its intrinsic value. This is because prior to expiration there is uncertainty in the value of the underlying and that gives the option some additional value.

The **time value** (or **time premium**) of an option is defined by the formula

$$\text{market price} = \text{intrinsic value} + \text{time value}$$

The time value represents the value that the option currently possesses due to the chance that its value will rise in the future: It is the *cost of risk*. The time value erodes to 0 as the expiration date approaches.

Figure 1.5 shows the time premiums for various times to expiration for a long call.

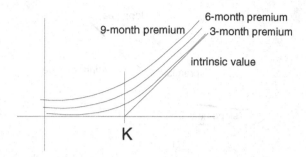

Figure 1.5: The time value of a long call

The time value of an option is small when the option is either far in or far out of the money. After all, an option that is far out of the money is likely to expire worthless and therefore does not have much potential value. Similarly, the final payoff of an option that is far in the money is very predictable, since its value varies roughly the same amount as the value of the underlying. Hence, there is not much time value in such an option. However, an option that is near the money has the potential of producing a significant percentage return, should the stock rise even slightly. This gives it a significant time value.

Note that the time value of an option is the reason that American options are seldom exercised early. After all, exercising an option yields a payoff equal to the intrinsic value, whereas sale of the option yields a payoff equal to the market value.

If an option is selling for its intrinsic value; that is, if the time value is 0, then the option is said to be selling **at parity**. Actually, sometimes a call option, especially one with a very high strike price, will trade *below* parity by a small amount. However, commissions generally negate the available profit for the average investor and so only certain investors (market-makers), whose commissions are very low, are in a position to profit from such options.

The Delta of an Option

There are several quantities associated with an option that measure how the option premium changes as some other quantity changes. These quantities are referred to as **Greeks**. One such Greek is the **delta**, which is the rate of change of the option premium with respect to the price of the underlying. The delta is thus the slope of the tangent line to the graph of the option premium, as shown in Figure 1.6 for a long call.

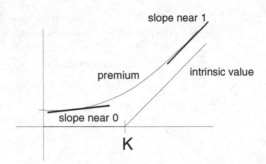

Figure 1.6: Delta

As can be seen from the figure, when the call is far in the money, a change in the underlying will produce an approximately equal change in the value of the call and so the delta is close to 1. On the other hand, if the call is far out of the money, a change in the underlying price will make little difference to the value of the option and so the delta is close to 0.

Another relevant Greek is called the **beta**. This is a measure of the change in the underlying price with respect to the market in general. A large beta indicates an underlying stock whose price is highly volatile; that is, subject to large rapid fluctuations. Stocks with large betas generally have more expensive options, because there is a greater chance that such options will become valuable (but also a greater chance that they will become worthless).

Selling Short

Short selling a stock is pictured in Figure 1.7 and proceeds as follows. Suppose an investor (the **short seller**) wants to short 100 shares of stock ABC. The investor requests the short sale from his broker. The broker locates a buyer for the stock and must also locate 100 shares of the stock, either in its own inventory, in one of its other client's accounts or at another institution. (Brokerage firms often have the right to borrow stocks held by their clients in a margin account.) The stock is borrowed from the lender and sold to the buyer. The proceeds of the sale are credited to the *short seller's* account. However, some brokers do not allow the funds to be withdrawn or to collect interest.

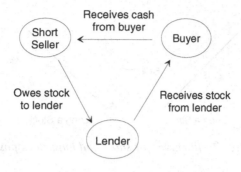

Figure 1.7

For this privilege, the short seller must return *the stock* to the lender upon demand. The short seller can **close** the short position, also called **unwinding** the short position at any time by purchasing the shares and returning them to the lender. The buyer of the shares is generally indifferent to these proceedings, unaware that his shares came from a short sale. The lender is also indifferent to the short sale. Indeed, the lender can still sell his stock even though it is technically part of a short sale, since the broker generally has sufficient shares to deal with this situation.

If the stock being shorted pays a dividend after the short sale, the *buyer* of the stock, who is the owner of record, will receive that dividend. On the other hand, the lender of the shares is also entitled to the dividend. Therefore, the *short seller must pay an amount of money equal to the dividend to the lender*.

As an aside, the practice of **naked short selling** is one in which the short seller receives the proceeds of a *fictious* sale of stock that is never actually located, borrowed or sold. The sale remains "open" until the seller (or broker) eventually does borrow the shares. This practice is illegal in the United States, but does seem to occur nonetheless.

As with short calls, selling a stock short incurs a potentially unlimited downside, unless the seller also owns shares of the stock with which to cover the inevitable return of the borrowed stock. Figure 1.8 shows the profit curve for a short sale of stock, as well as the profit curve for a long position.

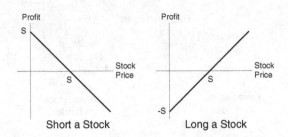

Figure 1.8: Profit curves for short and long stock positions

Exercises

1. Without looking in the book, draw the profit curves for a long put, short put, long call and short call.
2. Draw the payoff graph for the following option portfolio

$$-P_{80} + P_{100} + 2C_{130} - C_{150}$$

3. Describe an option portfolio that produces the payoff curve in the following figure, ignoring costs. (Assume that you can buy any number of options).

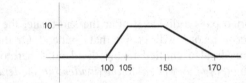

A **spread** is a transaction in which an investor simultaneously buys one option and sells another option, both on the same underlying asset, but with different terms (strike price and/or expiration date). A **call spread** involves the purchase and sale of calls, and similarly for a **put spread**. The idea is that one option is used to hedge the risk of the other option.

4. In a **bull call spread**, the investor buys a call at a certain strike price K_1 and sells another call at a higher strike price K_2, with the same expiration date. Draw the profit curve for a bull spread. When is a bull spread most profitable? Is this an optimistic or pessimistic investment? *Hint*: you must first decide how the costs of the two calls compare.
5. In a **bear call spread**, the investor buys a call at a certain strike price K_1 and sells another call at a lower strike price K_2, with the same expiration date. Draw the profit curve for a bear spread. When is a bear spread most profitable? Is this an optimistic or pessimistic investment? *Hint*: you must first decide how the costs of the two calls compare.
6. In a **calendar spread** also called a **time spread** an investor sells a call with a certain expiration date D_1 and buys a **more distant** call; that is, a call with a longer expiration date $D_2 > D_1$. Assume that the calls have the same

strike price. Consider the following calendar spread. The current (JAN) price of XYZ is $50. Call prices are as follows:

APR 50 call: (expiring in April at a strike price of $50) costs $5
JUL 50 call: $8
OCT 50 call: $10

Suppose that in 3 months (in April) the stock price is still $50. Then if all things else are equal the call prices should be

APR 50 call: $0 (expiring)
JUL 50 call: $5
OCT 50 call: $8

Is there a profit here for the investor? Describe the reason.

7. A **butterfly spread** is a combination of a bull spread and a bear spread. A call butterfly spread consists of buying a call at strike price K_1, selling two calls at strike price $K_2 > K_1$ and buying another call at strike price $K_3 > K_2$. All calls have the same expiration date. Draw a profit curve for a butterfly spread. *Hint*: They don't call it a butterfly spread for nothing.

Chapter 2
An Aperitif on Arbitrage

An **arbitrage opportunity** is an investment opportunity that is guaranteed not to result in a loss and *may* (with positive probability) result in a gain. We have remarked that if an arbitrage opportunity exists, then prices will be adjusted to eliminate that opportunity. This leads to the fundamental:

No-arbitrage Pricing Principle: *As a consequence of the tendency to an arbitrage-free market equilibrium, it only makes sense to price assets under the assumption that there is no arbitrage.* □

Forward Contracts

To give a simple example of how the no-arbitrage pricing principle is used to price assets, we discuss the pricing of forward contracts, beginning with some background on forward contracts.

Forward Contracts

A **forward contract** is an agreement to buy a certain quantity of an asset, called the **underlying asset** at a given price K, called the **settlement price** or **delivery price** to be paid at a given time T in the future, called the **settlement date** or **delivery date**. Entering a new forward contract does not require any initial purchase price—it is free.

The party that agrees to buy the asset is taking the **long position** on the contract and is said to be the **buyer** of the contract. The party that agrees to sell the asset is taking the **short position** on the contract and is said to be the **seller** of the contract.

Forward Prices and Delivery Prices

We have said that there is no cost to enter into a *new* forward contract. However, for an *existing* contract, there may be a fee (or a credit).

Consider a forward contract with delivery time T. At any time $t < T$, one can potentially enter into such a contract. The delivery price of the contract was fixed when the contract was first made, but as market prices fluctuate, the price that buyers will be willing to pay for the underlying for delivery at time T will probably change. Put another way, the delivery price for a *new* contract may be different from the delivery price of the older contract. The **forward price** $F_{t,T}$ at time t is the price for a *new* contract, established at time t. Thus, an existing forward contract whose delivery price is different from the current (time-t) forward price $F_{t,T}$ will have a nonzero value.

For example, suppose that on July 1 the delivery price of a contract to deliver 5000 bushels of wheat on October 1 is \$1.70 per bushel. Suppose that on August 1 the delivery price of a new forward contract for delivery on October 1 is \$1.80 per bushel. Then the original contract is worth $\$0.10 \times 5000 = \$500$.

Spot Prices

In contrast to forward prices, the **spot price** S_t of an asset at a given time t is the price of the asset *at that time* for *immediate* delivery. For example, we can speak of the current spot price of a bushel of wheat. We can also speak of the spot price of wheat in one month. This is the price that investors would pay in one month for immediate delivery at that time. Of course, at the present time, this spot price is unknown.

The Pricing of Forward Contracts

To determine the forward price of a forward contract, we can use a simple no-arbitrage argument. First, we need a definition. To **cash-and-carry** an asset means to borrow the cost of the asset and buy that asset. The resulting portfolio consists of the asset along with a debt equal to the initial cost of that asset. Note that the initial value of the portfolio is 0, since the two components balance each other out at the time of purchase. Thus, establishing a cash-and-carry portfolio is free. Of course, as time passes, the value of the portfolio will generally change. In particular, the debt will grow at the risk-free rate and the value of the asset will generally change as well.

The short position that corresponds to cash-and-carry is to borrow the asset, sell it and invest the resulting revenue at the risk-free rate. This position is referred to as **reverse cash-and-carry**. In the case of an option, however, reverse cash-and-carry involves simply selling the call and investing the proceeds (no need to borrow a call). Thus,

Cash-and-carry = Borrow money, hold asset
Reverse cash-and-carry = Borrow and sell asset, hold money

Now suppose that a forward contract is for an underlying that has initial spot price S_0.

In a perfect market, it is possible to go long or short on either the forward contract or on a cash-and-carry portfolio on the underlying asset. The final payoffs at the delivery time T for the long positions are

$$V_T(\text{long forward contract}) = S_T - F_{0,T}$$
$$V_T(\text{cash-and-carry the asset}) = S_T - S_0 e^{rT}$$

For example, in the case of cash-and-carry, at time T the investor owns the asset, worth S_T but must repay the loan with interest, at a cost of $S_0 e^{rT}$ where r is the risk-free rate. Of course, the final payoffs for the short positions are the negatives of the payoffs for the corresponding long positions.

Now, if an investor shorts the forward contract and cash-and-carries the asset, the initial cost is 0 and the final payoff is

$$V_T = (F_{0,T} - S_T) + (S_T - S_0 e^{rT}) = F_{0,T} - S_0 e^{rT}$$

which is a *known constant at time* 0; that is, it does not depend on the spot price S_T of the asset at a future time. Hence, it must be 0 or else there will be an arbitrage opportunity, for if $V_T > 0$, then this portfolio provides a guaranteed profit and if $V_T < 0$, then the reverse portfolio (long the forward contract and reverse cash-and-carry the asset) provides a guaranteed profit. Hence, the absence of arbitrage demands that the forward contract be priced at

$$F_{0,T} = S_0 e^{rT}$$

Theorem 2.1 *In a perfect market with no arbitrage, a forward contract to buy an asset with initial spot price S_0 at delivery time T has forward price*

$$F_{0,T} = S_0 e^{rT}$$

where r is the risk-free interest rate.□

Futures Contracts

In contrast to plain-vanilla forward contracts as described above, a **futures contract** is a forward contract with a number of constraints and a much more complicated payoff model. The main properties of futures contracts are as follows.

1) Futures contracts trade on an organized exchange. For example, the Chicago Board of Trade (CBT or CBOT) is the largest futures exchange.
2) Futures contracts have standardized terms, specifying the amount and precise type of the underlying, the delivery date and the delivery price.
3) As with stock options, performance (delivery of losses or gains) of futures contracts is guaranteed by a *clearinghouse*.
4) The purchase of a futures contract requires that the buyer post *margin*; that is, some amount of money to cover potential day-to-day price changes.

5) Futures markets are regulated by a government agency, whereas forward contracts are largely unregulated.
6) Futures contracts can be *closed* (terminated) either by delivery, by offset (that is, by a reversing trade that cancels both contracts) or by exchange-for-physical (which is a form of "settle up early" arrangement).

The forward price $f_{t,T}$ of a futures contract is called the **futures price** of the contract.

Daily Settlement

Unlike forward contracts, which are settled at the end of the contract, futures contracts undergo **daily settlement**, also called **daily marking-to-market**. For **end-of-day settlement**, at the end of each day's trading, the difference d in the futures prices at that time and at the close of the previous day's trading is computed. If $d > 0$, then the owner of the futures contract has actually made d dollars, since the contract would be worth d dollars to someone who wanted to buy such a forward contract. Hence, the amount d is credited to the owner's account. Similarly, if $d < 0$, then $|d|$ is debited from the owner's account. The reverse process applies to the *seller* of the futures contract.

Through the mechanism of daily settlement, at the end of any day after the start of a contract (say time 0), the buyer and seller of a futures contract are on the hook only for an amount equal to the *original* futures price $f_{0,T}$. Hence, the futures contract itself takes care of protecting the brokerage house for this commitment.

Since daily settlement provides a payment stream to the investor, the time value of money comes into play for a futures contract. However, if the risk-free rate is a known constant r throughout the model, then a simple no-arbitrage argument can be used to show that the *forward price* $F_{0,T}$ and the *futures price* $f_{0,T}$ for the same underlying with the same settlement date must be equal. For readability in the upcoming computation, we drop the subscript notation and write $F(0,T)$ in place of $F_{0,T}$ and $f(0,T)$ in place of $f_{0,T}$.

Theorem 2.2 *If the risk-free rate is a known constant r throughout the model, then the absence of arbitrage implies that the forward price $F(0,T)$ and the futures price $f(0,T)$ must be equal; that is, marking to market has no effect on the price of a contract to purchase in the future.*
Proof. We consider two portfolios, one involving simply a forward contract and the other involving futures contracts. The final value of the first portfolio; that is, of a forward contract entered at time $t = t_0$ is

$$\mathcal{V}_1 = S_T - F(0,T)$$

where S_T is the spot price of the underlying at time T.

For the portfolio of futures contracts, we will let the mark-to-market credits or debits grow at the risk-free rate and also adjust the amount of the futures contract held for each day by buying and selling portions of the contract, as can be done under the assumption of a perfect market.

Let t_i be the end-of-day settlement time for the ith day, where $i = 1, \ldots, N$. For the moment, we denote the quantity of futures contracts held on the ith day by q_i. We will determine the precise value of q_i in a few moments. At time t_i, the mark-to-market adjustment is

$$q_i[f(t_i, t_N) - f(t_{i-1}, t_N)]$$

where $t_0 = 0$ and $t_N = T$. This grows at time t_N to

$$q_i e^{r(t_N - t_i)}[f(t_i, t_N) - f(t_{i-1}, t_N)]$$

Adding these quantities gives a final value of

$$V_2 = \sum_{i=1}^{N} q_i e^{r(t_N - t_i)}[f(t_i, t_N) - f(t_{i-1}, t_N)]$$

Now, if we take $q_i = e^{-r(t_N - t_i)}$, then this sum collapses to

$$V_2 = f(t_N, t_N) - f(t_0, t_N) = S_T - f(0, T)$$

Note that the difference in the final values of the two portfolios is the *constant*

$$V_2 - V_1 = F(0, T) - f(0, T)$$

that is, it does not depend on future spot prices. Hence, since both portfolios have the same initial value (namely 0), the no-arbitrage principle implies that $V_2 - V_1 = 0$; that is,

$$F(0, T) = f(0, T) \qquad \qquad \square$$

The Put-Call Option Parity Formula

We can also apply the no-arbitrage pricing principle to derive relationships between the market prices of puts and calls in the same series; that is, with the same underlying, the same strike price and the same expiration date. The *put-call option parity formula* is a formula that compares the price P of a put to the price C of a call from the same series.

The European Case

Suppose that a stock is currently selling at a price of S_0 per share. A European put on this stock sells for P dollars and a European call sells for C dollars. Both options having the same strike price K and expiration time T. Suppose also that the present value of any dividends paid by the stock during the period in question is d_0. Let r denote the risk-free rate.

A portfolio consisting of one long European call and one short European put has initial value $C - P$ and time-T payoff

$$(S_T - K)^+ - (K - S_T)^+ = S_T - K$$

To duplicate this final payoff using stock, consider a portfolio consisting of a share of the stock and a debt of x dollars. This portfolio has initial value $S_0 - x$ and final payoff $S_T - xe^{rT} + d_0 e^{rT}$. We want the final payoffs to be the same; that is, we want

$$S_T - xe^{rT} + d_0 e^{rT} = S_T - K$$

and so $x = Ke^{-rT} + d_0$. Since the final payoffs are now the same, the absence of arbitrage implies that the initial values must be the same; that is,

$$C - P = S_0 - Ke^{-rT} - d_0$$

This is the put-call option parity formula.

Theorem 2.3 (European Options with Dividends) *Suppose that a stock is currently selling at a price of S_0 per share. A European put on this stock sells for P dollars and a European call for C dollars, both having the same strike price K and expiration time T. Suppose that the present value of any dividends paid by the stock during the period in question is d_0. Then the no-arbitrage pricing principle implies that*

$$C - P = S_0 - Ke^{-rT} - d_0$$

where r is the risk-free interest rate. This formula is called the **put-call option parity formula** *for European options.* □

The American Case

For the case of American options, we can only get inequalities. We use the same notation as in the European case. Let Portfolio A consist of a long American call and a short American put. The initial value is

$$\mathcal{V}_{A,0} = C - P$$

As to the final value, if the put is never exercised (that is, if it expires worthless), it is because the stock price S_T is greater than the strike price K and so the final value of the portfolio is the value $S_T - K$ of the call. If the put is exercised at time t, forcing us to buy the stock for K dollars and if we let the call expire, then the final value is

$$\mathcal{V}_{A,T} = S_T - Ke^{r(T-t)}$$

Since we have no control over the time t, the best we can say is that

$$S_T - Ke^{rT} \leq V_{A,T} \leq S_T - K$$

Now, let Portfolio B consist of the stock along with a debt of Ke^{-rT}. The initial value is

$$V_{B,0} = S_0 - Ke^{-rT}$$

and the final value is

$$V_{B,T} = S_T - K + d_0 e^{rT}$$

Therefore, since $V_{A,T} \leq V_{B,T}$, the no-arbitrage assumption implies that the initial costs must satisfy $V_{A,0} \leq V_{B,0}$; that is,

$$C - P \leq S_0 - Ke^{-rT}$$

This is one of the put-call option parity inequalities for American options.

For the other inequality, let Portfolio C consist of the stock along with a debt of $K + d_0$. The initial value is

$$V_{C,0} = S_0 - K - d_0$$

and the final value is

$$V_{C,T} = S_T - Ke^{rT} - d_0 e^{rT} + d_0 e^{rT} = S_T - Ke^{rT}$$

and since $V_{C,T} \leq V_{A,T}$, we have $V_{C,0} \leq V_{A,0}$; that is,

$$S_0 - K - d_0 \leq C - P$$

This is the other put-call option parity inequality for American options. We can now sumamrize.

Theorem 2.4 (American Options with Dividends) *Suppose that a stock is currently selling for S_0 per share, an American put on this stock sells for P and an American call sells for C, both having the same strike price K and expiration time T. The present value of any dividends paid by the stock during the period in question is d_0. Then assuming that no arbitrage occurs, we have*

$$S_0 - K - d_0 \leq C - P \leq S_0 - Ke^{-rT}$$

where r is the risk-free interest rate. This is called the **put-call option parity formula** *for American options.* $\square$

Comparing Option Prices

Since an American option provides all of the features of a corresponding European option and more, it seems obvious that American options should not be less expensive than their European counterparts. In symbols,

$$C^A \geq C^E, \quad P^A \geq P^E$$

It is not hard to see that it is possible for the price of an American put to exceed the price of its European counterpart. The idea is that early exercise of the American put can turn a share of stock into a certain amount of risk-free asset, which grows at the risk-free rate r. Therefore, if that rate is sufficiently high, the profit can be higher than that of the European put, which is limited by the strike price K (when the stock price plummets to 0).

Specifically, exercising an American put at time t and investing the resulting $K - S_t$ dollars at the risk-free rate r produces a profit of $(K - S_t)e^{r(T-t)}$ and so if

$$(K - S_t)e^{r(T-t)} > K$$

that is, if the time-t stock price satisfies

$$S_t < K(1 - e^{-r(T-t)})$$

then this plan is guaranteed to produce a greater profit than that of the corresponding European put.

On the other hand, it is a perhaps somewhat surprising fact that it is *never* advantageous to exercise an American call *on a nondividend paying stock* before expiration.

To see this, suppose first that the call is exercised at time $t < T$ with the intention of retaining the stock until time T. Then it would be preferable to wait until time T to exercise the call, since the cost K is the same in both cases but it is better to pay that cost at the later time.

On the other hand, suppose that the call is exercised at time $t < T$ with the intention of selling the stock. An alternative is to short the stock at time t. Then at time T, the call is exercised and the stock is used to unwind the short position. In the first case, the investor receives $S_t - K$ dollars at time t. In the second case, the investor receives S_t dollars at time t and pays K dollars at time T. Once again, it is better to pay the fixed amount K at the later time.

Exercises

1. If the underlying asset of a forward contract provides a dollar income during the life of the contract, then the long investor in the contract will lose out on this income and the cash-and-carry investor will get the income. This affects the previous no-arbitrage argument. In this situation, show that

$$F_{0,T} = (S_0 - I)e^{rT}$$

where I is the present value of the income.

2. We have seen that in the simplest case of a forward contract that does not produce an income, the nonarbitrage forward price at time 0 is

$$F_{0,T} = S_0 e^{rT}$$

However, we derived this formula under the very idealistic assumption of a perfect market. Let us examine what happens if this restriction is lifted. In particular, suppose that the lending and borrowing rates are different, as is almost always the case in real life. Let the lending rate for the investor be r_ℓ and the borrowing rate be r_b, where $r_\ell < r_b$. Show that the assumption of no arbitrage implies that

$$S_0 e^{r_\ell T} \le F_{0,T} \le S_0 e^{r_b T}$$

Hint: To avoid arbitrage, both strategies must yield a *nonpositive* payoff. The upper and lower bounds given in this exercise are called **no-arbitrage bounds** and the range of values of the future price that is implied by the absence of arbitrage is the **no-arbitrage spread**. Thus, in the absence of a perfect market, the lack of arbitrage implies that the future price can lie anywhere within a *range* of values.

3. Prove by an arbitrage argument that the initial value of a European or American call is less than the initial price of the stock; that is,

$$C^E \le S_0 \quad \text{and} \quad C^A \le S_0$$

4. Prove the following by an arbitrage argument

$$P^E \le K e^{-rT} \quad \text{and} \quad P^A \le K$$

5. Prove that

$$K e^{-rT} + d_0 - P^E \le S_0 \le K e^{-rT} + d_0 + C^E$$

6. Prove that

$$S_0 - K \le C^A$$
$$K - S_0 \le P^A$$

7. Prove that if a stock pays a dividend whose present value is d_0 then

$$K e^{-rT} + d_0 - S_0 \le P$$

where P is the price of a put.

Part 2—Discrete-Time Pricing Models

Chapter 3
Discrete Probability

Financial asset pricing involves the prediction of future events and as such relies very heavily on the mathematical theory of probability. In this chapter and the next, we describe the basic probability required for the study of discrete-time pricing models.

Partitions

Our discussion of pricing models will also rely heavily on the concept of a *partition* of a set, so we begin this chapter with a detailed look at partitions, after which we will return to the subject of probability.

Definition *Let Ω be a nonempty set. Then a **partition** of Ω is a collection $\mathcal{P} = \{B_1, \ldots, B_n\}$ of nonempty subsets of Ω, called the **blocks** of the partition, with the following properties:*
1) The blocks are pairwise disjoint; that is,

$$B_i \cap B_j = \emptyset$$

for all $i \neq j$.
2) The union of the blocks is all of Ω; that is,

$$B_1 \cup \cdots \cup B_n = \Omega$$

We denote the set of all partitions of X by Part(X).$\square$

Figure 3.1 shows a partition of a set Ω.

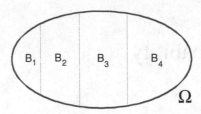

Figure 3.1: A partition of Ω

Refinements

If we further divide some of the blocks of a partition into smaller blocks, the resulting partition is called a *refinement* of the original partition. Here is the formal definition.

Definition *Let $\mathcal{P} = \{B_1, \ldots, B_k\}$ be a partition of a set Ω. Then a partition $\mathcal{Q} = \{C_1, \ldots, C_n\}$ is called a **refinement** of $\mathcal{P}$, written $\mathcal{P} \succ \mathcal{Q}$, if each block C_i of $\mathcal{Q}$ is completely contained in some block B_j of $\mathcal{P}$ or, equivalently, if each block of $\mathcal{P}$ is a union of blocks of $\mathcal{Q}$.* $\square$

Note that according to the definition a partition $\mathcal{P}$ is a refinement of itself. All other refinements of $\mathcal{P}$ are formed simply by breaking up one or more blocks of $\mathcal{P}$ into smaller blocks.

Figure 3.2 shows a refinement $\mathcal{Q} = \{C_1, \ldots, C_7\}$ of the partition in Figure 3.1. Note that

$$B_1 = C_1 \cup C_2$$
$$B_2 = C_3 \cup C_4 \cup C_5$$
$$B_3 = C_6$$
$$B_4 = C_7$$

and so each block of $\mathcal{Q}$ is completely contained in some block of $\mathcal{P}$.

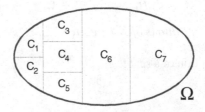

Figure 3.2: A refinement of the partition in Figure 3.1

The relation of refinement is actually a *partial order* on the family of all partitions of a set Ω. Specifically, we have the following, whose proof is left to the reader.

Theorem 3.1 *Let* Ω *be a nonempty set. The relation of refinement is a partial order on* Part(Ω); *that is, it satisfies the following properties.*

1)   (**Reflexive property**) *Every partition is a refinement of itself; that is,*

$$\mathcal{P} \succ \mathcal{P}$$

   for all $\mathcal{P} \in$ Part(Ω).

2)   (**Antisymmetric property**) *For all* $\mathcal{P}, \mathcal{Q} \in$ Part(Ω),

$$\mathcal{P} \succ \mathcal{Q} \quad and \quad \mathcal{Q} \succ \mathcal{P} \quad \Rightarrow \quad \mathcal{P} = \mathcal{Q}$$

3)   (**Transitive property**) *For all* $\mathcal{P}, \mathcal{Q}, \mathcal{R} \in$ Part(Ω),

$$\mathcal{P} \succ \mathcal{Q} \quad and \quad \mathcal{Q} \succ \mathcal{R} \quad \Rightarrow \quad \mathcal{P} \succ \mathcal{R} \qquad \square$$

Algebras

As mentioned, partitions play a very important role in the study of discrete-time pricing models. An equivalent concept is that of an *algebra*. Although we will not use algebras directly in our analysis of discrete-time pricing models, we want to discuss them very briefly here because they provide a helpful bridge to σ-*algebras*, which we will need for the analysis of continuous-time pricing models later in the book.

Definition *Let* Ω *be a nonempty set. A collection* $\mathcal{A}$ *of subsets of* Ω *is called an* **algebra of sets** *(or just an* **algebra***) if it satisfies the following properties:*
1)   *(Empty set is in* $\mathcal{A}$)

$$\emptyset \in \mathcal{A}$$

2)   *(*$\mathcal{A}$ *is closed under complements)*

$$A \in \mathcal{A} \quad \Rightarrow \quad A^c \in \mathcal{A}$$

3)   *(*$\mathcal{A}$ *is closed under unions)*

$$A, B \in \mathcal{A} \quad \Rightarrow \quad A \cup B \in \mathcal{A} \qquad \square$$

It is not hard to show that any algebra of sets is closed under intersections and set differences as well; that is,

$$A, B \in \mathcal{A} \quad \Rightarrow \quad A \cap B \in \mathcal{A} \quad and \quad A \setminus B \in \mathcal{A}$$

(Recall that $A \setminus B$ is the set of all elements of A that are *not* in B.) Also, the entire set Ω is a member of the algebra, since $\Omega = \emptyset^c$. The following concept makes precise the notion of the "smallest" nonempty sets in an algebra $\mathcal{A}$.

Definition *Let* $\mathcal{A}$ *be an algebra on a set* Ω. *An* **atom** *of* $\mathcal{A}$ *is a nonempty set* $S \in \mathcal{A}$ *with the property that no nonempty proper subset of* S *is also in* $\mathcal{A}$. *We denote the set of all atoms of* $\mathcal{A}$ *by* Atoms($\mathcal{A}$).$\square$

It seems reasonable that every member of an algebra on a *finite* set Ω is a union of atoms of $\mathcal{A}$. Here is the proof.

Theorem 3.2 *Let $\mathcal{A}$ be an algebra on a finite set Ω.*
1) *Every $\omega \in \Omega$ is contained in some atom of $\mathcal{A}$.*
2) *If A is an atom of $\mathcal{A}$ and $B \in \mathcal{A}$, then exactly one of the following must hold:*

$$A \cap B = \emptyset \quad or \quad A \subseteq B$$

  In words, every atom of $\mathcal{A}$ is either disjoint from B or contained in B.
3) *Every $B \in \mathcal{A}$ is a union of atoms of $\mathcal{A}$.*

Proof. For part 1), as pictured in Figure 3.3, let I be the intersection of all members of $\mathcal{A}$ that contain ω. (There is at least one member of $\mathcal{A}$ containing ω, namely, Ω itself.)

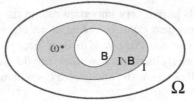

Figure 3.3

Then $\omega \in I \in \mathcal{A}$. To see that I is an atom, suppose not. Then there is a $B \in \mathcal{A}$ that is a *nonempty proper* subset of I. But $\omega \notin B$ since otherwise B would have been part of the intersection that defined I and so $I \subseteq B$, which contradicts $B \subset I$. But $\omega \in I \setminus B \in \mathcal{A}$ and since $I \setminus B$ is strictly smaller than I, we have a contradiction. Thus, I is an atom containing ω.

For part 2), the intersection $A \cap B \in \mathcal{A}$ is contained in the atom A and so $A \cap B = \emptyset$ or $A \cap B = A$; that is, $A \cap B = \emptyset$ or $A \subseteq B$. For part 3), part 1) implies that every $\omega \in B$ is contained in some atom $A_\omega \subseteq B$. Hence, part 2) implies that

$$B = \bigcup_{\omega \in B} A_\omega$$

that is, B is the union of the atoms. $\square$

Partitions and Algebras

For finite sample spaces, the notions of partition and algebra are equivalent. In particular, starting with a partition $\mathcal{P}$ of Ω, we can generate an algebra $\mathcal{A}(\mathcal{P})$ of sets simply by taking the empty set along with all possible unions of the blocks of $\mathcal{P}$. The reverse is also possible: Starting with an algebra of sets on a finite sample space Ω, the family of all atoms of $\mathcal{P}$ is a partition of Ω.

Theorem 3.3 *Let Ω be a nonempty finite set.*

1) *For any partition $\mathcal{P}$ of Ω the set* Unions($\mathcal{P}$) *consisting of the empty set along with all possible unions of the blocks of $\mathcal{P}$ is an algebra, called the* **algebra generated by** $\mathcal{P}$.

2) *If $\mathcal{A}$ is an algebra on Ω, then the set* Atoms($\mathcal{A}$) *of all atoms of $\mathcal{A}$ is a partition of Ω, called the* **partition defined by** $\mathcal{A}$.

Proof. We leave proof of part 1) to the reader. For part 2), let $\mathcal{A}$ be an algebra. To see that Atoms($\mathcal{A}$) is a partition of Ω, note first that atoms are nonempty by definition. Also, part 2) of Theorem 3.2 implies that distinct atoms are disjoint and part 1) of that theorem implies that the union of all atom of $\mathcal{A}$ is Ω. Thus, Atoms($\mathcal{A}$) is a partition of Ω.$\square$

Theorem 3.3 defines correspondences

$$\mathcal{P} \to \text{Unions}(\mathcal{P}) \quad \text{and} \quad \mathcal{A} \to \text{Atoms}(\mathcal{A})$$

from partitions to algebras and from algebras to partitions. In fact, these correspondences are inverses of each other (and are therefore one-to-one).

To see this, suppose that $\mathcal{P}$ is a partition of Ω. Then the smallest members of the algebra Unions($\mathcal{P}$) are certainly the blocks of $\mathcal{P}$ themselves; that is,

$$\text{Atoms}(\text{Unions}(\mathcal{P})) = \mathcal{P}$$

(An individual block $A \in \mathcal{P}$ is a "union" of blocks of $\mathcal{P}$, namely the union of the family $\{A\}$.) On the other hand, if $\mathcal{A}$ is an algebra on Ω, then all elements of $\mathcal{A}$ are unions of atoms of $\mathcal{A}$; that is,

$$\text{Unions}(\text{Atoms}(\mathcal{A})) = \mathcal{A}$$

This shows that the two correspondences are one-to-one and are inverses of each other.

The next theorem strengthens the connection between partitions and algebras: It says that the concept of refinement of partitions corresponds to set inclusion of algebras.

Theorem 3.4 *If $\mathcal{P}$ and $\mathcal{Q}$ are partitions of Ω, then*

$$\mathcal{P} \succ \mathcal{Q} \quad \Leftrightarrow \quad \text{Unions}(\mathcal{P}) \subseteq \text{Unions}(\mathcal{Q})$$

Proof. To say that $\mathcal{P} \succ \mathcal{Q}$ is to say that every block of $\mathcal{P}$ is a union of blocks of $\mathcal{Q}$; that is,

$$\mathcal{P} \subseteq \text{Unions}(\mathcal{Q})$$

But this is equivalent to

$$\text{Unions}(\mathcal{P}) \subseteq \text{Unions}(\mathcal{Q}) \qquad\qquad \square$$

Overview of Probability

We can now turn to the subject of probability.

Probability seems to have had its origins in an effort to predict the outcome of games of chance and is generally considered to have begun as a formal theory in a series of letters between the two famous mathematicians Blaise Pascal and Pierre de Fermat in the summer of 1654.

In the study of probability, the typical scenario is that of an *experiment*, such as rolling a pair of dice, administering a drug to a patient and recording a vital statistic or predicting the future price of a stock. The key is that the experiment must have a well-defined set of *possible outcomes*. This set is referred to as the **sample space** of the experiment.

When the sample space is a finite set, any subset of the sample space is referred to as an **event**. When an outcome occurs that is in a particular event, we say that the event *has occurred*. Thus, for example, we have the event of getting a sum of 7 on the dice, the event that a patient's systolic blood pressure drops to 110 after receiving a drug or the event that a stock price rises by 10%.

Next, a method must be determined to measure the *probability*, or *likelihood*, that various events will occur as a result of conducting the experiment. More specifically, the probability of an event is a real number between 0 and 1 that measures the likelihood that the outcome will lie in the event. A probability of 0 indicates that the event cannot occur (is impossible) and a probability of 1 indicates that the event is certain to occur.

The method that is used to determine these probabilities is not really part of the subject of probability per se. Two approaches are common. One is simply to assume the probabilities. For instance, consider the experiment of tossing a single coin. Assuming that the coin is fair is equivalent to assuming that the probabilities of heads and tails are both $1/2$. These probabilities are referred to as **theoretical probabilities**. Another approach is statistical in nature, using empirical data to assign probabilities. For example, if a coin is flipped 10000 times and results in 5003 heads, we may decide to set the probability of heads equal to $5003/10000$. This probability is referred to as an **empirical probability**.

The flavor of probability theory depends quite markedly on the size of the sample space. The basic concepts of probability theory require far less mathematical machinery when dealing with *finite* sample spaces, for in this case probabilities can be assigned to *individual outcomes* in the sample space. As we will soon see, all that is required is that the probabilities be numbers between 0 and 1 (inclusive) that add up to 1. Then the probability of an event is simply the sum of the probabilities of the outcomes that lie in that event. The term *finite probability theory* is used to refer to the theory of probability on finite sample spaces.

As an example, suppose that, based on market research, we decide that a certain stock, currently selling at $100 per share, will be selling at either $99, $100 or $101 by the end of the day. Thus, we have an experiment whose sample space consists of the possible stock prices

$$\Omega = \{99, 100, 101\}$$

Further, after research into the price history of the stock, we may decide to assign empirical probabilities as follows:

$$\mathbb{P}(99) = 0.25, \mathbb{P}(100) = 0.5, \mathbb{P}(101) = 0.25$$

In this case, the event that the price does not fall is $\{100, 101\}$ and its probability is $\mathbb{P}(100) + \mathbb{P}(101) = 0.75$.

For countably infinite sample spaces, probabilities can also be assigned to the individual outcomes in the sample space. However, the issue of convergence of an infinite sum now comes into play. The term *discrete probability* is used to refer to the probability of finite or countably infinite sample spaces.

As an example of a discrete (but nonfinite) sample space, consider the experiment of tossing a coin until the *first* heads appears and assume that the probability of heads on any given toss is p. The outcome is the toss number of this first heads. At the outset, we cannot confine the set of outcomes to any finite sample space, because there is no way to tell in advance how many tosses will be necessary before a heads appears. So the sample space must be the infinite set

$$\Omega = \{1, 2, 3, \dots\}$$

of all positive integers. Indeed, one must argue (or assume) that a heads must eventually appear, for if not then even this set does not represent all possible outcomes.

In this case, the probability that the so-called *waiting time* to the first heads is k is given by

$$\mathbb{P}(\text{first heads at toss } k) = (1-p)^{k-1}p$$

since the probability of tails is $1 - p$. Note that

$$\sum_{k \geq 1}(1-p)^{k-1}p = p\sum_{k \geq 0}(1-p)^k = p\frac{1}{1-(1-p)} = 1$$

The mathematics required for the study of probability on uncountable sample spaces (such as the real line) is decidedly more sophisticated. For example, imagine the stock in a company that is headed for bankruptcy. It is only a matter of time before the stock price is essentially 0. Let us call this the *time to failure* of the stock. The waiting time for this event could, at least in theory, be any positive real number (assume the stock trades 24 hours per day) so the sample space is the set Ω of all positive real numbers, which is uncountable.

However, unlike the case of a discrete sample space, we cannot assign a positive probability to each of uncountably many times to failure because the sum of *uncountably* many positive numbers is never finite, let alone equal to 1. So we must be content with assigning probabilities to certain "measurable" subsets of the sample space, which are referred to as **events**. Thus, in general, when the sample space of an experiment is uncountable, not every subset can qualify as an event; that is, not every subset can be assigned a probability.

The most direct and elegant way to assign probabilities to events is to use a *function*. Figure 3.4 shows how this might be done.

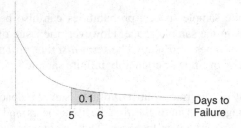

Figure 3.4: A probability density function

This figure shows the graph of a function that specifies the probability of failure for any *time interval*. In particular, it is the *area under the curve* that specifies the probability. For example, the probability that failure will occur sometime between the 5th and 6th day is the area under the curve between the vertical lines $x = 5$ and $x = 6$, which is 0.1. This function is referred to as a *probability density function*. Density functions are often used to specify probabilities in the nondiscrete case. However, there are some rather complex probability measures that cannot even be specified using a probability density function.

We will need only *finite* probability for our study of discrete-time pricing models. So let us proceed to set down the basic principles of the subject of finite probability. Since this is not a textbook on probability, we will tend to be brief, covering just what we need for our purposes. We will discuss nondiscrete (continuous) probability theory in Chapter 9.

Probability Spaces

We begin with the definition of a probability space.

Definition *A* **finite probability space** *is a pair* $(\Omega, \mathbb{P})$ *consisting of a finite nonempty set* Ω, *called the* **sample space** *and a real-valued function* $\mathbb{P}$ *defined on the set of all subsets of* Ω, *called a* **probability measure** *on* Ω. *The function* $\mathbb{P}$ *must satisfy the following properties.*

1) *(Range) For all* $A \subseteq \Omega$

$$0 \le \mathbb{P}(A) \le 1$$

2) *(Probability of* Ω*)*

$$\mathbb{P}(\Omega) = 1$$

3) **(Additivity property)** *If A and B are disjoint, then*

$$\mathbb{P}(A \cup B) = \mathbb{P}(A) + \mathbb{P}(B)$$

In this context, all subsets of Ω *are called* **events**. *For each* $\omega \in \Omega$ *the event* $\{\omega\}$ *is called an* **elementary event**. *A probability measure* $\mathbb{P}$ *is* **strongly positive** *if* $\mathbb{P}(A) > 0$ *for all nonempty events A.*□

As mentioned earlier, the sample space is intended to represent the set of all possible outcomes of an experiment and the probability $\mathbb{P}(\{\omega\})$ represents the likelihood that the outcome of the experiment will be ω. On the other hand, this is all intuition, not mathematics. Formally speaking, all we care about is that the pair $(\Omega, \mathbb{P})$ satisfies the definition given above.

Probability Mass Functions

The simplest way to define a probability measure on a finite sample space Ω is just to specify the probability of all elementary events. Equivalently, we assign to each of the elements $\omega \in \Omega$ a number p_ω satisfying $0 \le p_\omega \le 1$ and for which

$$\sum_{\omega \in \Omega} p_\omega = 1$$

Then we can define a probability measure $\mathbb{P}$ by setting

$$\mathbb{P}(\{\omega\}) = p_\omega$$

and extending this to all events by the additivity property; that is,

$$\mathbb{P}(A) = \sum_{\omega \in A} \mathbb{P}(\{\omega\})$$

The set $\{p_\omega \mid \omega \in \Omega\}$ is referred to as a **probability mass** or **probability distribution** and the function $f: \Omega \to \mathbb{R}$ defined by

$$f(\omega) = p_\omega$$

is called a **probability mass function**. (Do not confuse the term *probability distribution* as it is used here with the term *distribution function*, which has a different meaning that we will define in Chapter 9.)

It is commonplace to denote the probability $\mathbb{P}(\{\omega\})$ simply by $\mathbb{P}(\omega)$ and we shall do so as well. We emphasize that a probability *measure* is defined on all subsets of the (finite) sample space, whereas a probability *distribution* is defined only on the individual elements of the sample space.

Note that if each elementary event in Ω is equally likely; that is, if each outcome has the same probability, then this probability is $1/|\Omega|$ and so the probability of any event A is simply the size of A divided by the size of the sample space Ω; that is

$$\mathbb{P}(A) = \frac{|A|}{|\Omega|}$$

Example 3.1 Studies of the price history of a certain stock over the last several years have shown that, for the month of January, the probability that the stock will reach a certain maximum value during the month is as follows:

$$\mathbb{P}(0-4.99) = 0.65$$
$$\mathbb{P}(5-9.99) = 0.2$$
$$\mathbb{P}(10-14.99) = 0.1$$
$$\mathbb{P}(15-19.99) = 0.04$$
$$\mathbb{P}(20-24.99) = 0.01$$

What is the probability that the stock will reach \$10 during the month? What is the probability that the stock will be less than \$5 or at least \$20 sometime during the month?

Solution The stock will reach \$10 during the month if and only if the maximum stock price during the month is at least 10. Hence

$$\mathbb{P}(\text{price reaches 10})$$
$$= \mathbb{P}(10-14.99) + \mathbb{P}(15-19.99) + \mathbb{P}(20-24.99)$$
$$= 0.1 + 0.04 + 0.01 = 0.15$$

Similarly,

$$\mathbb{P}(\text{price} < 5 \text{ or } \geq 20) = \mathbb{P}(0-4.99) + \mathbb{P}(20-24)$$
$$= 0.65 + 0.01 = 0.66$$

$\square$

Probability theory tends to have its own vocabulary, even when it comes to simple concepts like the disjointness of sets.

Definition *When two events A and B are disjoint as sets; that is, when $A \cap B = \emptyset$, we say that they are* **mutually exclusive.** *When a collection $\{A_1, \ldots, A_n\}$ of events satisfies*

$$A_i \cap A_j = \emptyset$$

for all $i \neq j$, we say that the collection is **pairwise mutually exclusive.** $\square$

Some easy consequences of the definition of probability space are given below.

Theorem 3.5 *Let $(\Omega, \mathbb{P})$ be a finite probability space. Then*
1) *(Probability of the empty event)*

$$\mathbb{P}(\emptyset) = 0$$

2) *(Monotonicity) For any events A and B,*

$$A \subseteq B \quad \Rightarrow \quad \mathbb{P}(A) \leq \mathbb{P}(B)$$

3) *(Probability of the complement) For any event A,*

$$\mathbb{P}(A^c) = 1 - \mathbb{P}(A)$$

4) **(Finite additivity)** *If the events $A_1, \ldots, A_n$ are pairwise mutually exclusive, then*

$$\mathbb{P}(A_1 \cup \cdots \cup A_n) = \mathbb{P}(A_1) + \cdots + \mathbb{P}(A_n)$$ $\square$

The Theorem on Total Probability

The following important theorem says that we can determine the probability of an event E if we can determine the probability of that portion of E that belongs to each block of a partition of the sample space Ω.

Theorem 3.6 (Theorem on Total Probability) *Let Ω be a sample space and let $\{E_1, \ldots, E_n\}$ be a partition of Ω. Then for any event A in Ω,*

$$\mathbb{P}(A) = \sum_{k=1}^{n} \mathbb{P}(A \cap E_k)$$

Proof. Since the events E_k are mutually disjoint, so are the events $A \cap E_k$ and so finite additivity implies that the sum on the right above is equal to the probability of the union of the sets $A \cup E_k$; that is,

$$\sum_{k=1}^{n} \mathbb{P}(A \cap E_k) = \mathbb{P}\left(\bigcup_{k=1}^{n} (A \cup E_k) \right)$$

But

$$\bigcup_{k=1}^{n} (A \cup E_k) = A \cup \bigcup_{k=1}^{n} E_k = A \cup \Omega = A$$ $\square$

Independence

Many experiments involve the repetition of a simpler experiment under identical conditions. For example, we may toss a coin n times or draw a card from a deck of cards n times, replacing the previously drawn card each time. The mathematical concept that corresponds to the notion of *identical conditions* is *independence*.

Intuitively speaking, two events are independent if the knowledge (or assumption) that one event will happen does not affect the probability of the other event happening. Thus, for example, if we toss a coin twice, the tosses are independent if knowledge that the first toss resulted in heads (or tails) does not affect the probability that the second toss results in heads (or tails). Thus, for a sequence of 100 independent tosses of a fair coin, if the first 99 tosses result in heads (admittedly a very unlikely event), then the probability that the 100th toss is heads is still $1/2$. Many amateur gamblers would do well to remember this fact.

We will be able to make the statement about previous knowledge precise when we discuss conditional probability in the next chapter. For now, we can state the formal definition of independence as follows.

Definition *The events E and F on the probability space $(\Omega, \mathbb{P})$ are* **independent** *if the probability that both events occur is the product of the probability of each event; that is, if*

$$\mathbb{P}(E \cap F) = \mathbb{P}(E)\mathbb{P}(F) \qquad \square$$

For example, suppose that a certain stock can move up or down in price over a day and a certain bond can do likewise. If we *assume* that the actions of the stock and the bond are independent, then

$$\mathbb{P}(\text{stock up } and \text{ bond down}) = \mathbb{P}(\text{stock up})\mathbb{P}(\text{bond down})$$

We can also define independence of a collection of events.

Definition *The collection of events $\{E_1, \dots, E_n\}$ is* **independent** *if for every subcollection $\{E_{i_1}, \dots, E_{i_k}\}$ consisting of at least two of these events, we have*

$$\mathbb{P}(E_{i_1} \cap \cdots \cap E_{i_k}) = \mathbb{P}(E_{i_1}) \cdots \mathbb{P}(E_{i_k}) \qquad \square$$

Note that to check whether or not 3 events A, B and C are independent, we must check 4 conditions:

$$\mathbb{P}(A \cap B) = \mathbb{P}(A)\mathbb{P}(B)$$
$$\mathbb{P}(A \cap C) = \mathbb{P}(A)\mathbb{P}(C)$$
$$\mathbb{P}(B \cap C) = \mathbb{P}(B)\mathbb{P}(C)$$
$$\mathbb{P}(A \cap B \cap C) = \mathbb{P}(A)\mathbb{P}(B)\mathbb{P}(C)$$

In general, to check that a collection of k events is independent, we must check a total of $2^k - k - 1$ conditions, since this is the number of subcollections of the k events that contain at least two events. Thus, the number of conditions grows very rapidly with the number of events.

We will also have reason to consider families of collections of events; for example,

$$\mathcal{C}_1 = \{E_1, E_2, E_3\}, \quad \mathcal{C}_2 = \{E_4, E_5\}, \quad \mathcal{C}_3 = \{E_6\}$$

is a family of 3 collections of events. We can extend the definition of independence to such families as follows.

Definition *The collections $\mathcal{C}_1, \ldots, \mathcal{C}_k$ of events are* **independent** *if for every choice of one event $E_i \in \mathcal{C}_i$ from each collection, the events $E_1, \ldots, E_k$ are independent.* $\square$

The Binomial Distribution

The simplest type of meaningful experiment is one that has only two outcomes. Such experiments are referred to as **Bernoulli experiments**, or **Bernoulli trials**. The two outcomes are often described by the terms **success** and **failure** and the probability of success is often denoted by p and the probability of failure by $q = 1 - p$.

For example, tossing a coin is a Bernoulli experiment, where we may consider heads as success and tails as failure. As a more relevant example, we will consider a derivative pricing model in which at any given time t_k, the price of a certain stock may rise from its previous value S to Su or it may fall from its previous value S to Sd, where $0 < d < 1 < u$. Thus, at each time t_k we have a Bernoulli experiment.

If a Bernoulli experiment with probability of success p is repeated n times under identical conditions, this is called a **binomial experiment** with n **trials**. The sample space for this experiment is the set $\Omega = \{s, f\}^n$ of all sequences of s's and f's of length n, where s stands for success and f for failure. The **parameters** of the binomial experiment are the probability p of success and the number n of trials.

Assuming that the conditions are identical is equivalent to assuming that the outcomes of the trials are *independent*; that is, the outcome of the first k trials does not affect the outcome of the $(k + 1)$-st trial.

For example, tossing a coin n times is a binomial experiment. Drawing a card n times, where success is the drawing of an ace, is a binomial experiment *provided that we replace each card before drawing the next card*. This is necessary since we must repeat the *same* binomial experiment each time.

The assumption that the individual Bernoulli trials in a binomial experiment are independent leads us to define a probability distribution on $\Omega = \{s, f\}^n$ by

$$\mathbb{P}(\omega) = p^{N_s(\omega)} q^{n - N_s(\omega)}$$

where $q = 1 - p$ and where $N_s(\omega)$ is the number of successes in ω. For instance,

$$\mathbb{P}(ssfsf) = p^3 q^2$$

To see that this defines a probability distribution, first we observe that

$$0 \le \mathbb{P}(\omega) \le 1$$

for any $\omega \in \{s, f\}^n$.

To see that the sum of the probabilities is 1, note that the outcomes with exactly k successes are the sequences of s's and f's of length n that have exactly k s's. But the number of such sequences is the same as the number of ways to choose k positions from n for the s's and this is the binomial coefficient $\binom{n}{k}$. Hence, the probability of having exactly k successes is

$$\binom{n}{k} p^k q^{n-k}$$

and so by grouping according to the number of successes, we get

$$\sum_{\omega \in \Omega} \mathbb{P}(\omega) = \sum_{k=0}^{n} \binom{n}{k} p^k q^{n-k} = (p + q)^n = 1$$

Thus, $\mathbb{P}$ is a probability measure and so defines a probability space $(\Omega, \mathbb{P})$.

Theorem 3.7 *Consider a binomial experiment with sample space $\Omega = \{s, f\}^n$ and parameters p and n. Let $q = 1 - p$. For any $\omega \in \Omega$, let*

$$N_s(\omega) = number\ of\ s's\ in\ \omega$$

1) The function

$$\mathbb{P}(\omega) = p^{N_s(\omega)} q^{n - N_s(\omega)}$$

is a probability distribution on Ω.

2) *The probability of getting* exactly k *successes is*

$$\mathbb{P}(exactly\ k\ successes) = \binom{n}{k}p^k q^{n-k} \qquad \square$$

Example 3.2 Consider a stock whose price can change at any one of 6 times

$$t_0 < t_1 < \cdots < t_5$$

Suppose the stock's initial price at time t_0 is S. During each time interval $[t_k, t_{k+1}]$, the stock price goes up by a factor of u or down by a factor of d, where $0 < d < 1 < u$, independently of the previous changes in the price. The probability that the stock price goes up is p. Thus, for each time interval we have a Bernoulli experiment with probability of success p and the entire price history is a binomial experiment with parameters p and $n = 5$.

A typical outcome of this experiment can be written as a sequence of U's and D's of length 5 and so the sample space is the set

$$\Omega = \{U, D\}^5$$

The probability of having exactly 3 up-ticks in the stock price is thus

$$\binom{5}{3}p^3 q^2 = 10p^3 q^2$$

The probability of having *at least* 3 up-ticks in the stock price is the sum of the probabilities of having exactly 3, 4 or 5 successes, which is

$$\binom{5}{3}p^3 q^2 + \binom{5}{4}p^4 q + \binom{5}{5}p^5 = 10p^3 q^2 + 5p^4 q + p^5$$

so if $p = 1/2$, then this probability is

$$10\left(\frac{1}{2}\right)^3\left(\frac{1}{2}\right)^2 + 5\left(\frac{1}{2}\right)^4\left(\frac{1}{2}\right) + \left(\frac{1}{2}\right)^5 = 16\left(\frac{1}{2}\right)^5 = \frac{1}{2} \qquad \square$$

The probability distribution described in the previous example and theorem is extremely important.

Definition *Let* $0 < p < 1$ *and let* n *be a positive integer. Let* $\Omega = \{0, \ldots, n\}$. *The probability distribution on* Ω *with mass function*

$$b(k; n, p) = \binom{n}{k}p^k(1 - p)^{n-k}$$

for $k = 0, \ldots, n$ *is called the* **binomial distribution**. *This distribution gives the probability of getting exactly* k *successes in a binomial experiment with parameters* p *and* n. $\square$

Figure 3.5 shows the graph of two binomial distributions.

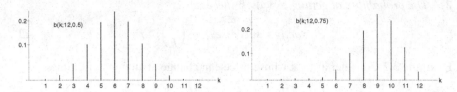

Figure 3.5: Binomial distributions

Conditional Probability

It is often the case that after a probability space $(\Omega, \mathbb{P})$ has been established, additional information becomes available about the possible outcomes of the experiment under consideration. This information comes in the form of knowledge that future outcomes must lie in an event E. This new information can be used to define a *new* probability measure on Ω, denoted by $\mathbb{P}(\ \cdot\ \mid E)$ and called a *conditional probability measure*.

Definition *Let* $(\Omega, \mathbb{P})$ *be a probability space. Let* E *be an event with* $\mathbb{P}(E) > 0$. *Then for any event* A, *the* **conditional probability of** A **given** E *is defined by*

$$\mathbb{P}(A \mid E) = \frac{\mathbb{P}(A \cap E)}{\mathbb{P}(E)} \qquad \square$$

The symbol $\mathbb{P}(A \mid E)$ is read "the probability of A given E." We do not define conditional probability when $\mathbb{P}(E) = 0$, for it makes little sense to ask about a probability conditioned upon the occurrence of an impossible event.

Example 3.3 Consider the experiment of drawing two cards at random from a deck of cards without replacing the first card before drawing the second card. What is the probability of drawing a spade on the second card given that the first card is a heart? What is the probability of drawing a spade on the second card given that the first card is a spade?
Solution The sample space consists of the set of all ordered pairs (c_1, c_2), where c_1 is the first card drawn, c_2 is the second card drawn and $c_1 \neq c_2$. Let S be the event of drawing a spade on the second card and let H be the event of drawing a heart on the first card. We seek the conditional probability

$$\mathbb{P}(S \mid H) = \frac{\mathbb{P}(S \cap H)}{\mathbb{P}(H)}$$

Now, $\mathbb{P}(H) = 13/52 = 1/4$ and

$$\mathbb{P}(S \cap H) = \frac{13 \times 13}{52 \times 51} = \frac{1}{4} \times \frac{13}{51}$$

and so

$$\mathbb{P}(S \mid H) = \frac{13}{51}$$

For the second problem, let A be the event of drawing a space on the first card. Then

$$\mathbb{P}(S \mid A) = \frac{\mathbb{P}(S \cap A)}{\mathbb{P}(A)} = \frac{13 \times 12}{52 \times 51} \bigg/ \frac{1}{4} = \frac{12}{51} \qquad \Box$$

Conditioning on an event defines a new "conditional" probability measure on Ω.

Theorem 3.8 *Let $(\Omega, \mathbb{P})$ be a finite probability space and let E be an event for which $\mathbb{P}(E) > 0$. Then the function $\mathbb{P}(\cdot \mid E)$ defined above is a probability measure on Ω for which $\mathbb{P}(E \mid E) = 1$.*
Proof. Monotonicity of $\mathbb{P}$ implies that

$$0 \le \mathbb{P}(A \cap E) \le \mathbb{P}(E)$$

and so

$$0 \le \mathbb{P}(A \mid E) \le 1$$

Also

$$\mathbb{P}(\Omega \mid E) = \frac{\mathbb{P}(\Omega \cap E)}{\mathbb{P}(E)} = \frac{\mathbb{P}(E)}{\mathbb{P}(E)} = 1$$

Finally, if $A \cap B = \emptyset$ then since $A \cap E$ and $B \cap E$ are also disjoint, we have

$$\begin{aligned}
\mathbb{P}(A \cup B \mid E) &= \frac{\mathbb{P}((A \cup B) \cap E)}{\mathbb{P}(E)} \\
&= \frac{\mathbb{P}((A \cap E) \cup (B \cap E))}{\mathbb{P}(E)} \\
&= \frac{\mathbb{P}(A \cap E) + \mathbb{P}(B \cap E)}{\mathbb{P}(E)} \\
&= \frac{\mathbb{P}(A \cap E)}{\mathbb{P}(E)} + \frac{\mathbb{P}(B \cap E)}{\mathbb{P}(E)} \\
&= \mathbb{P}(A \mid E) + \mathbb{P}(B \mid E)
\end{aligned}$$

and so $\mathbb{P}(\cdot \mid E)$ is finitely additive.$\Box$

The theorem on total probability takes on a nice appearance using conditional probabilities.

Theorem 3.9 (Theorem on Total Probability) *Let Ω be a sample space and let $E_1, \ldots, E_n$ form a partition of Ω for which $\mathbb{P}(E_k) > 0$ for all k. Then for any* *event A in* Ω,

$$\mathbb{P}(A) = \sum_{k=1}^{n} \mathbb{P}(A \mid E_k)\mathbb{P}(E_k) \ . \qquad \square$$

Random Variables

In finite probability theory, real-valued functions also have a special name.

Definition *A real-valued function* $X : \Omega \to \mathbb{R}$ *defined on a finite sample space* Ω *is called a* **random variable** *on* Ω. *The set of all random variables on* Ω *is denoted by* $\mathrm{RV}(\Omega)$. $\square$

For finite probability spaces, a random variable is nothing more or less than a real-valued function. However, as we will see in Chapter 9, for nondiscrete sample spaces, not all real-valued functions qualify as random variables.

If X is a random variable, it is customary to denote the inverse image of a set B under X not by $X^{-1}(B)$, as for ordinary functions, but instead by

$$\{X \in B\}$$

Also, instead of writing $X^{-1}(x)$ it is customary to write

$$\{X = x\}$$

Example 3.4 Let

$$\Omega = \{0.5, 0.75, 1, 1.25, 1.5, 1.75\}$$

be a sample space of possible federal discount rates. (The federal **discount rate** is the interest rate that the Federal Reserve charges member banks to borrow money from the Federal government, through the so-called **discount window**.) Consider a company whose stock price tends to fluctuate with interest rates. The stock prices can be represented by a random variable S on Ω. For example

$$S(0.5) = 105$$
$$S(0.75) = 100$$
$$S(1) = 100$$
$$S(1.25) = 100$$
$$S(1.5) = 95$$
$$S(1.75) = 90$$

The event $\{S = 100\}$ is the event consisting of the discount rates $\{0.75, 1, 1.25\}$; that is,

$$\{S = 100\} = \{0.75, 1, 1.25\} \qquad \square$$

Example 3.5 Consider the experiment of rolling two fair dice and recording the values on each die. The sample space consists of the 36 ordered pairs

$$\Omega = \{(1,1),(1,2),(1,3),\ldots,(6,4),(6,5),(6,6)\}$$

Since the dice are fair, each ordered pair is equally likely to occur and so the probability of each outcome is $1/36$.

However, for some games of chance, we are interested only in the *sum* of the two numbers on the dice. So let us define a random variable $S\colon \Omega \to \mathbb{R}$ by

$$S(a,b) = a + b$$

The event $\{S = 7\}$ of getting a sum of 7 is

$$\{S = 7\} = \{(1,6),(2,5),(3,4),(4,3),(5,2),(6,1)\}$$

and

$$\mathbb{P}(\text{sum equals } 7) = \mathbb{P}(S = 7) = \frac{6}{36} = \frac{1}{6} \qquad \square$$

The set $\mathrm{RV}(\Omega)$ is a *vector space* under ordinary addition and scalar multiplication of functions. Thus, if X and Y are random variables on Ω and $a, b \in \mathbb{R}$ then

$$aX + bY$$

is a random variable on Ω. Also, the product XY of two random variables on Ω, defined by

$$(XY)(\alpha) = X(\alpha)Y(\alpha)$$

is a random variable on Ω.

Indicator Random Variables

As shown in the previous example, random variables are used to identify events, such as $\{X = x\}$ and $\{X \in B\}$. Given an event A, there is a random variable that identifies A specifically.

Definition *Let A be an event in Ω. The function 1_A defined by*

$$1_A(\omega) = \begin{cases} 1 & \omega \in A \\ 0 & \omega \notin A \end{cases}$$

is called the **indicator function** *or* **indicator random variable** *for A.* $\square$

We will have frequent use of the fact that

$$1_A 1_B = 1_{A \cap B}$$

and so, in particular, if A and B are disjoint, then

$$1_A 1_B = 0$$

The Partition of a Random Variable

If X is a random variable (i.e. function) on a finite sample space Ω, then the image $\text{im}(X)$ is a finite set, say

$$\text{im}(X) = \{x_1, \ldots, x_n\}$$

Moreover, the inverse images

$$\{X = x_1\}, \ldots, \{X = x_n\}$$

of the elements of $\text{im}(X)$ form a partition of Ω. Indeed, the block $\{X = x_i\}$ is the set of all elements $\omega \in \Omega$ that are sent to x_i by the function X.

Definition *Let X be a random variable on Ω with*

$$\text{im}(X) = \{x_1, \ldots, x_n\}$$

Then the partition

$$\mathcal{P}_X = \{\{X = x_1\}, \ldots, \{X = x_n\}\}$$

is called the **partition defined by** X.$\square$

The Probability Distribution of a Random Variable

If X is a random variable on Ω and if $\mathbb{P}$ is a probability measure on Ω, it is customary to denote the probability of the event $\{X = x\}$ by $\mathbb{P}(X = x)$; that is,

$$\mathbb{P}(X = x) = \mathbb{P}(\{X = x\})$$

The partition $\mathcal{P}_X$ defined by X then defines a probability measure $\mathbb{P}_X$ on $\text{im}(X) = \{x_1, \ldots, x_n\}$ by

$$\mathbb{P}_X(x_i) = \mathbb{P}(X = x_i)$$

for all $x_i \in \text{im}(X)$. This probability distribution on $\text{im}(X)$ is called the **probability distribution** of X and the corresponding probability mass function $f: \mathcal{A} \to \mathbb{R}$ defined by

$$f(x_i) = \mathbb{P}(X = x_i)$$

is called the **probability mass function** of X.

It is common to speak of the probability distribution of a random variable X without mentioning the sample space that is the domain of X. For example, to say that a random variable X has a binomial distribution with parameters p and n is to say that the values of X are $\{0, \ldots, n\}$ and that the probability mass function of X is the function

$$\mathbb{P}(X = k) = b(k; n, p) = \binom{n}{k} p^k (1 - p)^n$$

It is also common to say in this case that X is **binomially distributed**.

Measurability of a Random Variable with Respect to a Partition

Of course, a random variable X is *constant* on the blocks $\{X = x\}$ of its partition $\mathcal{P}_X$, since by definition, X takes the value x on the block $\{X = x\}$. This property has a special name.

Definition *Let $\mathcal{Q} = \{B_1, \ldots, B_n\}$ be a partition of Ω. A random variable X on Ω is $\mathcal{Q}$-measurable if X is constant on each block of $\mathcal{Q}$; that is, if it has the form*

$$X = \sum_{i=1}^{n} b_i 1_{B_i}$$

for (not necessarily distinct) constants $b_i \in \mathbb{R}$. $\square$

As just mentioned, a random variable X is measurable with respect to its own partition $\mathcal{P}_X$, since

$$X = \sum_{i=1}^{n} x_i 1_{\{X = x_i\}}$$

where $\operatorname{im}(X) = \{x_1, \ldots, x_n\}$. The thing that distinguishes $\mathcal{P}_X$ from all other partitions $\mathcal{Q}$ for which X is $\mathcal{Q}$-measurable is that X takes on a *different* constant value on each block of the partition $\mathcal{P}_X$; that is, the constants b_i in the definition above are distinct.

It follows easily that X is measurable with respect to any refinement $\mathcal{Q}$ of $\mathcal{P}_X$, since each block of $\mathcal{Q}$ is contained in a single block of $\mathcal{P}_X$. Conversely, if X is $\mathcal{Q}$-measurable for some partition $\mathcal{Q}$, then $\mathcal{Q}$ must be a refinement of $\mathcal{P}_X$, since if a block B of $\mathcal{Q}$ should intersect two distinct blocks of $\mathcal{P}_X$ nontrivially, then X would not be constant on B.

In summary, a random variable X is $\mathcal{Q}$-measurable if and only if $\mathcal{Q}$ is a refinement of $\mathcal{P}_X$, in symbols,

$$X \text{ is } \mathcal{Q}\text{-measurable} \quad \Leftrightarrow \quad \mathcal{P}_X \succ \mathcal{Q}$$

In fact, we can characterize the partition $\mathcal{P}_X$ in either of the following equivalent ways.

Theorem 3.10 *Let X be a random variable on Ω.*

1) $\mathcal{P}_X$ *is the only partition of* X *with the property that* X *is* $\mathcal{Q}$*-measurable for* $\mathcal{Q} \in \mathrm{Part}(\Omega)$ *if and only if* $\mathcal{P}_X \succ \mathcal{Q}$*.*

2) $\mathcal{P}_X$ *is the coarsest partition of* Ω *under which* X *is measurable; that is,* $\mathcal{P}_X$ *is the largest partition of* Ω *under the partial order of refinement on* $\mathrm{Part}(\Omega)$ *under which* X *is measurable.*$\square$

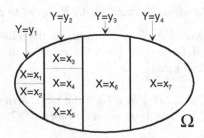

Figure 3.6

Let X and Y be random variables on Ω. It follows from Theorem 3.10 that

$$Y \text{ is } \mathcal{P}_X\text{-measurable} \quad \Leftrightarrow \quad \mathcal{P}_Y \succ \mathcal{P}_X$$

Moreover, if $\mathcal{P}_Y \succ \mathcal{P}_X$, then it is easy to see that Y must be a function of X. Loosely speaking, this follows from the fact that if we know the value $X(\omega)$, then we know which block $A \in \mathcal{P}_X$ contains ω and so we also know which block $B \in \mathcal{P}_Y$ contains ω. Hence, we know $Y(\omega)$.

In more detail (see Figure 3.6), each block $B \in \mathcal{P}_Y$ is the union of blocks of $\mathcal{P}_X$, say

$$B = A_1 \cup \cdots \cup A_k$$

and suppose that Y takes the value y on the elements of B and X takes the value a_i on the elements of A_i, for each $i = 1, \ldots, k$. Define a function f by $f(x_i) = y$ for all i. Then for any $\omega \in B$, we have

$$f(X(\omega)) = y = Y(\omega)$$

and so $Y = f(X)$.

For instance, referring to Figure 3.6, if the function f is defined by

$$\begin{aligned}
f(x_1) &= f(x_2) = y_1 \\
f(x_3) &= f(x_4) = f(x_5) = y_2 \\
f(x_6) &= y_4 \\
f(x_7) &= y_5
\end{aligned}$$

then

$$Y = f(X)$$

Theorem 3.11 *Let X and Y be random variables on Ω. The following are equivalent:*
1) Y *is* $\mathcal{P}_X$*-measurable*
2) $\mathcal{P}_Y \succ \mathcal{P}_X$
3) Y *is a function of* X*; that is, there is a function* $f \colon \mathbb{R} \to \mathbb{R}$ *for which*

$$Y = f(X)$$

Proof. We have seen that 1) and 2) are equivalent and have also proved that 1) implies 3). Finally, if $Y = f(X)$, then Y is equal to the constant $f(x)$ on the block $\{X = x\}$ of $\mathcal{P}_X$ and so Y is $\mathcal{P}_X$-measurable. Hence, 3) implies 1).$\square$

Independence of Random Variables

If X and Y are random variables on the same sample space Ω, then we write

$$\mathbb{P}(X = x, Y = y)$$

to mean $\mathbb{P}(\{X = x\} \cap \{Y = y\})$. By definition, X and Y are *independent* if the events $\{X = x\}$ and $\{Y = y\}$ are independent; that is, if

$$\mathbb{P}(X = x, Y = y) = \mathbb{P}(X = x)\mathbb{P}(Y = y)$$

for all choices of $x \in \mathrm{im}(X)$ and $y \in \mathrm{im}(Y)$.

Taking a cue from the definition of independence for three events, for the independence of three random variables X, Y and Z, we might expect the requirements that

1) $\mathbb{P}(X = x, Y = y) = \mathbb{P}(X = x)\mathbb{P}(Y = y)$
2) $\mathbb{P}(X = x, Z = z) = \mathbb{P}(X = x)\mathbb{P}(Z = z)$
3) $\mathbb{P}(Y = y, Z = z) = \mathbb{P}(Y = y)\mathbb{P}(Z = z)$
4) $\mathbb{P}(X = x, Y = y, Z = z) = \mathbb{P}(X = x)\mathbb{P}(Y = y)\mathbb{P}(Z = z)$

for all $x \in \mathrm{im}(X)$, $y \in \mathrm{im}(Y)$ and $z \in \mathrm{im}(Z)$. However, 4) actually implies 1)–3). For example, the theorem on total probability gives

$$
\begin{aligned}
\mathbb{P}(X = x, Y = y) &= \sum_{z \in \Omega} \mathbb{P}(X = x, Y = y, Z = z) \\
&= \sum_{z \in \Omega} \mathbb{P}(X = x)\mathbb{P}(Y = y)\mathbb{P}(Z = z) \\
&= \mathbb{P}(X = x)\mathbb{P}(Y = y) \sum_{z \in \Omega} \mathbb{P}(Z = z) \\
&= \mathbb{P}(X = x)\mathbb{P}(Y = y)
\end{aligned}
$$

which is 1). The others are proved similarly. Thus, the definition of independence of a family of random variables is much simpler than that of a family of events!

Definition

1) The random variables X and Y on Ω are **independent** *if*

$$\mathbb{P}(X = x, Y = y) = \mathbb{P}(X = x)\mathbb{P}(Y = y)$$

for all $x \in \text{im}(X)$ and $y \in \text{im}(Y)$.

2) More generally, the random variables $X_1, \ldots, X_n$ are independent if

$$\mathbb{P}(X_1 = x_1, \ldots, X_n = x_n) = \prod_{i=1}^{n} \mathbb{P}(X_i = x_i)$$

for all $x_i \in \text{im}(X_i)$. $\square$

To generalize our previous argument, if $X_1, \ldots, X_n$ are independent and if

$$\text{im}(X_n) = \{a_1, \ldots, a_m\}$$

then the theorem on total probability gives

$$\mathbb{P}(X_1 = x_1, \ldots, X_{n-1} = x_{n-1})$$

$$= \sum_{i=1}^{m} \mathbb{P}(X_1 = x_1, \ldots, X_{n-1} = x_{n-1}, X_n = a_i)$$

$$= \prod_{i=1}^{n-1} \mathbb{P}(X_i = x_i) \sum_{i=1}^{m} \mathbb{P}(X_n = a_i)$$

$$= \prod_{i=1}^{n-1} \mathbb{P}(X_i = x_i)$$

More generally, for any nonempty subset $\{i_1, \ldots, ., i_m\}$ of $\{1, \ldots, ., n\}$, we have

$$\mathbb{P}(X_{i_1} = x_{i_1}, \ldots, X_{i_m} = x_{i_m}) = \prod_{j=1}^{m} \mathbb{P}(X_{i_j} = x_{i_j})$$

which is why we did not need to explicitly include this condition in the definition of independence of random variables.

Positive Random Variables

In later chapters, we will need to identify the following types of random variables.

Definition *Let X be a random variable on Ω. Then*

1) X is **nonnegative**, *written $X \geq 0$ if*

$$X(\omega) \geq 0 \text{ for all } \omega \in \Omega$$

Note that the zero random variable $X = 0$ is nonnegative.

2) X is **strictly positive**, written $X > 0$ if $X \geq 0$ but $X \neq 0$; that is, if

$$X(\omega) \geq 0 \text{ for all } \omega \in \Omega \text{ and } X(\omega) > 0 \text{ for at least one } \omega \in \Omega$$

3) X is **strongly positive**, written $X \gg 0$ if

$$X(\omega) > 0 \text{ for all } \omega \in \Omega \qquad\qquad \square$$

Random Vectors

We will also need random *vectors*.

Definition *Let* $(\Omega, \mathbb{P})$ *be a probability space. A function* $X \colon \Omega \to \mathbb{R}^n$ *is called a* **random vector** *on* Ω. *We denote the set of all such random vectors by* $\mathrm{RV}^n(\Omega)$.$\square$

Of course, random variables are random vectors (take $n = 1$). The set $\mathrm{RV}^n(\Omega)$ is a *vector space* under ordinary addition and scalar multiplication of functions. We leave proof of this fact as an exercise.

Example 3.6 Let

$$\Omega = \{0.5, 0.75, 1, 1.25, 1.5, 1.75\}$$

be a sample space of possible federal discount rates. Consider a company whose stock price tends to fluctuate with interest rates. Of course, bond prices also fluctuate with respect to interest rates. We might define the price random vector $S \colon \Omega \to \mathbb{R}^2$ by $S(\omega) = (s, b)$ where s is the price of the stock and b is the price of the bond when the discount rate is ω. For example,

$$S(0.5) = (105, 112)$$

means that if the discount rate is 0.5% then the stock price is 105 and the bond price is 112.$\square$

Expectation

The notion of expected value plays a central role in the mathematics of finance.

Definition *Let* X *be a random variable on a finite probability space* $(\Omega, \mathbb{P})$. *The* **expected value** (*also called the* **expectation** *or* **mean**) *of* X *is given by*

$$\mathcal{E}_{\mathbb{P}}(X) = \sum_{\omega \in \Omega} X(\omega)\mathbb{P}(\omega)$$

In words, the expected value of X *is the weighted sum of the values of* X, *each weighted by a probability of occurrence. The expected value of* X *is also denoted by* μ_X.$\square$

When the probability measure is clear from the context, we will drop the subscript and write $\mathcal{E}$ for $\mathcal{E}_\mathbb{P}$. We invite the reader to prove the following.

Theorem 3.12 *If a random variable* $X: \Omega \to \mathbb{R}$ *takes on the* distinct *values* $\{x_1, \ldots, x_m\}$, *then*

$$\mathcal{E}(X) = \sum_{i=1}^{m} x_i \mathbb{P}(X = x_i) \qquad \square$$

The expected value function $\mathcal{E}: \mathrm{RV}(\Omega) \to \mathbb{R}$ maps random variables on $(\Omega, \mathbb{P})$ to real numbers. One of the most important properties of this function is that it is *linear*.

Theorem 3.13 *The expectation function* $\mathcal{E}: \mathrm{RV}(\Omega) \to \mathbb{R}$ *is a linear functional; that is, for any random variables X and Y and real numbers a and b,*

$$\mathcal{E}(aX + bY) = a\mathcal{E}(X) + b\mathcal{E}(Y)$$

Proof. Let us suppose that X has values $\{x_1, \ldots, x_n\}$ and Y has values $\{y_1, \ldots, y_m\}$. Then $aX + bY$ has values $ax_i + by_j$ for $i = 1, \ldots, n$ and $j = 1, \ldots, m$. Hence, the theorem on total probability gives

$$\mathcal{E}(aX + bY) = \sum_{i=1}^{n}\sum_{j=1}^{m} (ax_i + by_j)\mathbb{P}(X = x_i, Y = y_j)$$

$$= a\sum_{i=1}^{n} x_i \left[\sum_{j=1}^{m} \mathbb{P}(X = x_i, Y = y_j)\right]$$

$$+ b\sum_{j=1}^{m} y_j \left[\sum_{i=1}^{n} \mathbb{P}(X = x_i, Y = y_j)\right]$$

$$= a\sum_{i=1}^{n} x_i \mathbb{P}(X = x_i) + b\sum_{j=1}^{m} y_j \mathbb{P}(Y = y_j)$$

$$= a\mathcal{E}(X) + b\mathcal{E}(Y)$$

as desired. $\square$

Of course, the expected value $\mathcal{E}(X)$ of a random variable X is a constant. In fact, $\mathcal{E}(X)$ is the *best possible* approximation of X by a constant, as measured by the **mean squared error** or **MSE**, defined by

$$\mathrm{MSE}_c = \mathcal{E}[(X - c)^2]$$

where c is a constant. In particular, for all constants c, the mean squared error MSE_c is smallest if and only if $c = \mu_X$; that is,

$$\mathcal{E}[(X - \mu_X)^2] \le \mathcal{E}[(X - c)^2]$$

with equality if and only if $c = \mu_X$. To prove this, we note that $\mathcal{E}(X - \mu_X) = 0$ and so

$$
\begin{aligned}
\mathcal{E}[(X - c)^2] &= \mathcal{E}[\{(X - \mu_X) + (\mu_X - c)\}^2] \\
&= \mathcal{E}[(X - \mu_X)^2] + 2\mathcal{E}[(X - \mu_X)(\mu_X - c)] + \mathcal{E}[(\mu_X - c)^2] \\
&= \mathcal{E}[(X - \mu_X)^2] + \mathcal{E}[(\mu_X - c)^2] \\
&\geq \mathcal{E}[(X - \mu_X)^2]
\end{aligned}
$$

with equality holding if and only if $c = \mu_X$.

Theorem 3.14 *Let X be a random variable on a probability space $(\Omega, \mathbb{P})$. Then the expected value $\mu_X = \mathcal{E}(X)$ of X is the best constant approximation to X in the sense of mean squared error; that is,*

$$
\mathcal{E}[(X - \mu_X)^2] \leq \mathcal{E}[(X - c)^2]
$$

with equality if and only if $c = \mu_X$. $\square$

Expected Value of a Function of a Random Variable

Note that if $f: \mathbb{R} \to \mathbb{R}$ is a function and X is a random variable, then the composition $f(X): \Omega \to \mathbb{R}$ is also a random variable. (For finite probability spaces, this is nothing more than the statement that the composition of two functions is also a function.) The expected value of $f(X)$ is equal to

$$
\mathcal{E}(f(X)) = \sum_{i=1}^{n} f(X(\omega_i)) \mathbb{P}(\omega_i)
$$

If X takes on the distinct values $x_1, \ldots, x_m$, then

$$
\mathcal{E}(f(X)) = \sum_{i=1}^{m} f(x_i) \mathbb{P}(X = x_i)
$$

Example 3.7 Consider a stock whose current price is 100 and whose price at time T depends on the state of the economy, which may be one of the following states:

$$
\Omega = \{\mathfrak{s}_1, \mathfrak{s}_2, \mathfrak{s}_3, \mathfrak{s}_4\}
$$

The probabilities of the various states are given by

$$
\begin{aligned}
\mathbb{P}(\mathfrak{s}_1) &= 0.2 \\
\mathbb{P}(\mathfrak{s}_2) &= 0.3 \\
\mathbb{P}(\mathfrak{s}_3) &= 0.3 \\
\mathbb{P}(\mathfrak{s}_4) &= 0.2
\end{aligned}
$$

The stock price random variable is given by

$$S(\mathbf{s}_1) = 99$$
$$S(\mathbf{s}_2) = 100$$
$$S(\mathbf{s}_3) = 101$$
$$S(\mathbf{s}_4) = 102$$

If we purchase one share of the stock now the expected return at time T is

$$\mathcal{E}(S) = 99(0.2) + 100(0.3) + 101(0.3) + 102(0.2) = 100.5$$

and so the expected profit is $100.5 - 100 = 0.5$. Consider a derivative whose return D is a function of the stock price, say

$$D(99) = -4$$
$$D(100) = 5$$
$$D(101) = 5$$
$$D(102) = -6$$

Thus D is a random variable on Ω. The expected return of the derivative is

$$\begin{aligned}
\mathcal{E}(\text{return}) &= D(99)\mathbb{P}(99) + D(100)\mathbb{P}(100) \\
&\quad + D(101)\mathbb{P}(101) + D(102)\mathbb{P}(102) \\
&= -4(0.2) + 5(0.3) + 5(0.3) - 6(0.2) \\
&= 1
\end{aligned}$$

$\square$

The previous example points out a key property of expected values. The expected value is seldom the value expected! In this example, we never expect to get a return of 100.5. In fact, this return is impossible. The return must be one of the numbers in the sample space. The expected value is an *average*, not the value most expected.

Expectation and Independence

We have seen that the expected value operator is linear; that is,

$$\mathcal{E}(aX + bY) = a\mathcal{E}(X) + b\mathcal{E}(Y)$$

It is natural to wonder also about $\mathcal{E}(XY)$. Let us suppose that X has values $\{x_1, \ldots, x_n\}$ and Y has values $\{y_1, \ldots, y_m\}$. Then the product XY has values $x_i y_j$ for $i = 1, \ldots, n$ and $j = 1, \ldots, m$ and so

$$\mathcal{E}(XY) = \sum_{i=1}^{n} \sum_{j=1}^{m} x_i y_j \mathbb{P}(X = x_i, Y = y_j)$$

In general, we can simplify this no further. However, if X and Y are *independent*, then

$$\mathcal{E}(XY) = \sum_{i=1}^{n}\sum_{j=1}^{m} x_i y_j \mathbb{P}(X = x_i, Y = y_j)$$

$$= \sum_{i=1}^{n}\sum_{j=1}^{m} x_i y_j \mathbb{P}(X = x_i)\mathbb{P}(Y = y_j)$$

$$= \left[\sum_{i=1}^{n} x_i \mathbb{P}(X = x_i)\right]\left[\sum_{j=1}^{m} y_j \mathbb{P}(Y = y_j)\right]$$

$$= \mathcal{E}(X)\mathcal{E}(Y)$$

Thus, we have an important theorem.

Theorem 3.15 *If X and Y are independent random variables on a probability space $(\Omega, \mathbb{P})$ then*

$$\mathcal{E}(XY) = \mathcal{E}(X)\mathcal{E}(Y) \qquad \square$$

This theorem can be generalized to the product of more than two independent random variables. For example, if X, Y and Z are independent, then XY and Z are also independent and so

$$\mathcal{E}(XYZ) = \mathcal{E}(XY)\mathcal{E}(Z) = \mathcal{E}(X)\mathcal{E}(Y)\mathcal{E}(Z)$$

Variance and Standard Deviation

The expectation of a random variable X is a measure of the "center" of the distribution of X. A common measure of the "spread" of the values of a random variable is the variance and its square root, which is called the standard deviation. The advantage of the standard deviation is that it has the same units as the random variable. However, its disadvantage is the often awkward presence of the square root.

Definition *Let X be a random variable with finite expected value μ. The* **variance** *of X is*

$$\sigma_X^2 = \mathrm{Var}(X) = \mathcal{E}((X - \mu)^2)$$

and the **standard deviation** *is the positive square root of the variance*

$$\sigma_X = \sqrt{\mathrm{Var}(X)} \qquad \square$$

The following theorem gives some simple properties of the variance, the first of which is often quite useful in computing the variance. Proof is left as an exercise.

Theorem 3.16 *Let X be a random variable with finite expected value μ. Then*
1) $\mathrm{Var}(X) = \mathcal{E}(X^2) - \mu^2 = \mathcal{E}(X^2) - \mathcal{E}(X)^2$

2)   *For any real number a,*

$$\text{Var}(aX) = a^2\text{Var}(X)$$

3)   *If X and Y are* independent *random variables, then*

$$\text{Var}(X+Y) = \text{Var}(X) + \text{Var}(Y)$$

4)   *If c is a constant, then*

$$\text{Var}(X+c) = \text{Var}(X)$$ $\square$

Note that, unlike the expectation operator, the variance is *not* linear. Thus, in general,

$$\text{Var}(aX+bY) \neq a\text{Var}(X) + b\text{Var}(Y)$$

Standardizing a Random Variable

If X is a random variable with expected value μ and variance σ^2, then we can define a new random variable Y by

$$Y = \frac{X-\mu}{\sigma}$$

Then

$$\mathcal{E}(Y) = \frac{1}{\sigma}(\mathcal{E}(X) - \mu) = 0$$

and

$$\text{Var}(Y) = \frac{1}{\sigma^2}\text{Var}(X-\mu) = \frac{1}{\sigma^2}\text{Var}(X) = 1$$

Thus, Y has expected value 0 and variance 1. The process of passing from X to Y is called **standardizing** the random variable X and a random variable with mean 0 and variance 1 is called a **standard random variable**.

Expected Value of a Bernoulli Random Variable

A **Bernoulli random variable** is a random variable that takes only two possible values. Suppose that X is a Bernoulli random variable taking on the two values α and β with probabilities

$$\mathbb{P}(X=\alpha) = p \quad \text{and} \quad \mathbb{P}(X=\beta) = q$$

where $q = 1 - p$. Then

$$\mathcal{E}(X) = \alpha p + \beta q$$

and

$$\text{Var}(X) = \mathcal{E}(X^2) - \mathcal{E}(X)^2 = \alpha^2 p + \beta^2 q - (\alpha p + \beta q)^2$$

In the important special case $\alpha = 1$ and $\beta = 0$, we get

$$\mathcal{E}(X) = p \quad \text{and} \quad \text{Var}(X) = pq$$

We will have use for the following fact, whose proof is left as an exercise.

Theorem 3.17 *If X is a standard Bernoulli random variable with probabilities p and $q = 1 - p$, then X takes on the values*

$$\frac{q}{\sqrt{pq}} \quad \text{and} \quad \frac{-p}{\sqrt{pq}}$$

and

$$\mathbb{P}\left(X = \frac{q}{\sqrt{pq}}\right) = p \quad \text{and} \quad \mathbb{P}\left(X = \frac{-p}{\sqrt{pq}}\right) = q \qquad \square$$

Expected Value of a Binomial Random Variable

We can easily compute the expected value and variance of a binomial random variable.

Theorem 3.18 *Let X be a binomial random variable with distribution $b(k; n, p)$. Then*

$$\mathcal{E}(X) = np \quad \text{and} \quad \text{Var}(X) = npq$$

where $q = 1 - p$.
Proof. For the expected value, we have

$$
\begin{aligned}
\mathcal{E}(X) &= \sum_{k=0}^{n} k\mathbb{P}(X = k) \\
&= \sum_{k=1}^{n} k\binom{n}{k} p^k q^{n-k} \\
&= np\sum_{k=1}^{n} \binom{n-1}{k-1} p^{k-1} q^{(n-1)-(k-1)} \\
&= np\sum_{k=0}^{n-1} \binom{n-1}{k} p^k q^{(n-1)-k} \\
&= np
\end{aligned}
$$

Alternatively, X is the sum $B_1 + \cdots + B_n$ of independent Bernoulli random variables B_i, where $B_i = 1$ if the ith trial is a success and $B_i = 0$ if the ith trial is a failure. Hence, by the linearity of the expected value, we have

$$\mathcal{E}(X) = \sum_{i=1}^{n} \mathcal{E}(B_i) = \sum_{i=1}^{n} p = np$$

Moreover, since the random variables B_i are independent, we have

$$\text{Var}(X) = \sum_{i=1}^{n} \text{Var}(B_i) = \sum_{i=1}^{n} pq = npq \qquad \square$$

Conditional Expectation

We can put together the notions of conditional probability and expectation to get *conditional expectation*, which plays a key role in derivative pricing models.

Conditional Expectation with Respect to an Event

Conditional expectation with respect to an event A with positive probability is pretty straightforward—we just take the ordinary expectation but with respect to the conditional probability measure $\mathbb{P}(\cdot \mid A)$.

Definition *Let $(\Omega, \mathbb{P})$ be a finite probability space and let A be an event for which $\mathbb{P}(A) > 0$. The **conditional expectation** of a random variable X with respect to the event A is*

$$\mathcal{E}(X \mid A) = \sum_{\omega \in \Omega} X(\omega)\mathbb{P}(\omega \mid A)$$

The symbol $\mathcal{E}(X \mid A)$ is read the expected value of X given A. $\square$

Note that since the conditional expectation is just a special type of expectation, it is linear; that is,

$$\mathcal{E}(aX + bY \mid A) = a\mathcal{E}(X \mid A) + b\mathcal{E}(Y \mid A)$$

for random variables X and Y and real numbers a and b.

A little algebra gives a very useful expression for the conditional expectation in terms of the nonconditional expectation.

Theorem 3.19 *Let $(\Omega, \mathbb{P})$ be a finite probability space and let A be an event with $\mathbb{P}(A) > 0$. Then*

$$\mathcal{E}(X \mid A) = \frac{\mathcal{E}(X 1_A)}{\mathbb{P}(A)}$$

where 1_A is the indicator function of A.

Proof. In general, if $\omega \in \Omega$ and $A \subseteq \Omega$, then

$$\mathbb{P}(\omega \mid A) = \frac{\mathbb{P}(\{\omega\} \cap A)}{\mathbb{P}(A)} = \frac{\mathbb{P}(\omega)}{\mathbb{P}(A)} 1_A(\omega)$$

and so

$$\mathcal{E}(X \mid A) = \sum_{\omega \in \Omega} X(\omega) \mathbb{P}(\omega \mid A)$$

$$= \frac{1}{\mathbb{P}(A)} \sum_{\omega \in \Omega} X(\omega) 1_A(\omega) \mathbb{P}(\omega)$$

$$= \frac{1}{\mathbb{P}(A)} \mathcal{E}(X 1_A)$$

$\square$

One simple consequence of the previous theorem is the following useful result, whose proof we leave as an exercise.

Theorem 3.20 *If A and B are events with $\mathbb{P}(A \cap B) > 0$, then*

$$\mathcal{E}_{\mathbb{P}(\cdot \mid A)}(X \mid B) = \mathcal{E}(X \mid A \cap B)$$

$\square$

Next, we have the expected value analog of the theorem on total probability.

Theorem 3.21 *Let $\mathcal{P} = \{B_1, \ldots, B_n\}$ be a partition of Ω. Then for any random variable X on Ω*

$$\mathcal{E}(X) = \sum_{i=1}^{n} \mathcal{E}(X \mid B_i) \mathbb{P}(B_i)$$

Moreover, if $\mathbb{P}(A) > 0$, then

$$\mathcal{E}(X \mid A) = \sum_{i=1}^{n} \mathcal{E}(X \mid B_i \cap A) \mathbb{P}(B_i \mid A)$$

These sums are valid provided that we consider each term in which the conditional probability is not defined as equal to 0.

Proof. For the first part, since

$$X = \sum_{i=1}^{n} X 1_{B_i}$$

taking the expected value gives

$$\mathcal{E}(X) = \sum_{i=1}^{n} \mathcal{E}(X 1_{B_i}) = \sum_{i=1}^{n} \mathcal{E}(X \mid B_i) \mathbb{P}(B_i)$$

with the understanding that if $\mathbb{P}(B_i) = 0$, then $\mathcal{E}(X \mid B_i) \mathbb{P}(B_i) = 0$. For the second part, we apply the first part to the conditional probability $\mathbb{P}(\cdot \mid A)$,

$$\mathcal{E}(X \mid A) = \sum_{i=1}^{n} \mathcal{E}_{\mathbb{P}(\cdot \mid A)}(X \mid B_i)\mathbb{P}(B_i \mid A)$$

$$= \sum_{i=1}^{n} \mathcal{E}(X \mid B_i \cap A)\mathbb{P}(B_i \mid A)$$

where the undefined terms; that is, the terms for which $\mathbb{P}(B_i \cap A) = 0$, are 0.$\square$

Conditional Expectation with Respect to a Partition

Next we define conditional expectation with respect to a partition of the sample space. Unlike the conditional expectation with respect to an event, which is a real number, the conditional expectation given a partition is a *random variable*.

We have seen that the expected value $\mathcal{E}(X)$ of a random variable X is the best approximation of X by a *constant*, as measured by the mean squared error. Formally, for any constant c,

$$\mathcal{E}[(X - \mu_X)^2] \leq \mathcal{E}[(X - c)^2]$$

with equality if and only if $c = \mu_X$.

More generally, if $\mathcal{P} = \{B_1, \ldots, B_n\}$ is a partition of Ω for which each block B_i has positive probability, we wish to find the best approximation Y to X that is constant *on each block* of $\mathcal{P}$; that is, the best $\mathcal{P}$-*measurable* approximation to X. Thus, Y has the form

$$Y = \sum_{i=1}^{n} b_i 1_{B_i}$$

The quality of the approximation is measured by the mean squared error

$$\text{MSE} = \mathcal{E}[(X - Y)^2]$$

Since $X = X1_{B_1} + \cdots + X1_{B_n}$, we can write

$$\text{MSE} = \mathcal{E}[(X - Y)^2] = \mathcal{E}\left[\left\{\sum_{i=1}^{n}(X - b_i)1_{B_i}\right\}^2\right]$$

But $1_{B_i}1_{B_j} = 0$ if $i \neq j$ and so all cross terms in the expansion under the square brackets are equal to 0. Thus, since $(1_{B_i})^2 = 1_{B_i}$, we have

$$\text{MSE} = \mathcal{E}\left[\sum_{i=1}^{n}(X - b_i)^2 1_{B_i}\right] = \sum_{i=1}^{n}\mathcal{E}[(X - b_i)^2 1_{B_i}]$$

But each term above is nonnegative and so the MSE will be smallest when each term is smallest. But each term can be written in the form

$$\mathcal{E}[(X - b_i)^2 1_{B_i}] = \mathbb{P}(B_i)\mathcal{E}[(X - b_i)^2 \mid B_i]$$

The expression $\mathcal{E}[(X - b_i)^2 \mid B_i]$ is the mean squared error of X with respect to the (conditional) probability measure $\mathbb{P}(\cdot \mid B_i)$ and so we have seen that this is smallest when

$$b_i = \mathcal{E}_{\mathbb{P}(\cdot \mid B_i)}(X) = \mathcal{E}(X \mid B_i)$$

Hence, the best approximation of X among the $\mathcal{P}$-measurable random variables is given by

$$Y = \sum_{i=1}^{n} \mathcal{E}(X \mid B_i) 1_{B_i}$$

This leads to the following definition.

Definition *Let $(\Omega, \mathbb{P})$ be a finite probability space and let $\mathcal{P} = \{B_1, \ldots, B_n\}$ be a partition of Ω for which $\mathbb{P}(B_i) > 0$ for all i. The **conditional expectation** $\mathcal{E}(X \mid \mathcal{P})$ of a random variable X with respect to the partition $\mathcal{P}$ is the random variable*

$$\mathcal{E}(X \mid \mathcal{P}) = \mathcal{E}(X \mid B_1)1_{B_1} + \cdots + \mathcal{E}(X \mid B_n)1_{B_n}$$

In particular, for any $\omega \in \Omega$, if $\omega \in B_k$ then

$$\mathcal{E}(X \mid \mathcal{P})(\omega) = \mathcal{E}(X \mid B_k) \qquad \qquad \square$$

Note that for the partition $\mathcal{P} = \{\Omega\}$ that consists of the single block Ω, we have

$$\mathcal{E}(X \mid \{\Omega\}) = \mathcal{E}(X)$$

and so the ordinary expectation is a form of conditional expectation.

Figure 3.7 shows the best approximation for a random variable X that is defined on an interval $[a, b]$ of the real line. (We chose this illustration because it is easier to picture the conditional expectation for such intervals than for random variables on a finite sample space).

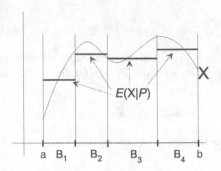

Figure 3.7: Conditional expectation

Here is a formal statement of the role of the conditional expectation in approximating X.

Theorem 3.22 *The random variable $\mathcal{E}(X \mid \mathcal{P})$ is the best approximation to X among all $\mathcal{P}$-measurable random variables, in the sense of mean squared error; that is,*

$$\mathcal{E}[(X - \mathcal{E}(X \mid \mathcal{P}))^2] \leq \mathcal{E}[(X - Y)^2]$$

for all $\mathcal{P}$-measurable random variables Y, with equality holding if and only if $Y = \mathcal{E}(X \mid \mathcal{P})$.□

The following theorem gives some key properties of conditional expectation. We will use them frequently.

Theorem 3.23 *Let $(\Omega, \mathbb{P})$ be a finite probability space. Let $\mathcal{P} = \{B_1, \ldots, B_n\}$ be a partition of Ω for which $\mathbb{P}(B_i) > 0$ for all i. The conditional expectation $\mathcal{E}(X \mid \mathcal{P})$ has the following properties.*

1) *The function $\mathcal{E}(\,\cdot\,\mid \mathcal{P})$ is linear; that is, for random variables X and Y and real numbers a and b,*

$$\mathcal{E}(aX + bY \mid \mathcal{P}) = a\mathcal{E}(X \mid \mathcal{P}) + b\mathcal{E}(Y \mid \mathcal{P}) \quad \bullet$$

2) *(Taking out what is known) If Y is a $\mathcal{P}$-measurable random variable, then*

$$\mathcal{E}(YX \mid \mathcal{P}) = Y\mathcal{E}(X \mid \mathcal{P})$$

In particular, if Y is $\mathcal{P}$-measurable then

$$\mathcal{E}(Y \mid \mathcal{P}) = Y$$

3) *(**The Tower Properties**) If $\mathcal{P} \succ \mathcal{Q}$, then*

$$\mathcal{E}(\mathcal{E}(X \mid \mathcal{P}) \mid \mathcal{Q}) = \mathcal{E}(X \mid \mathcal{P}) = \mathcal{E}(\mathcal{E}(X \mid \mathcal{Q}) \mid \mathcal{P})$$

In words, the expected value with respect to both $\mathcal{P}$ and $\mathcal{Q}$, taken in either order, is the same as the expected value with respect to the coarser of $\mathcal{P}$

and Q. In particular,

$$\mathcal{E}(\mathcal{E}(X \mid Q)) = \mathcal{E}(X)$$

4) *(An Independent Condition Drops Out) If X and $\mathcal{P}$ are independent; that is, if $\mathcal{P}_X$ and $\mathcal{P}$ are independent, then*

$$\mathcal{E}(X \mid \mathcal{P}) = \mathcal{E}(X)$$

Proof. Part 1) follows from the fact that the conditional expectation with respect to an event is linear. We leave details of a proof to the reader. To prove 2), since conditional expectation is linear and since Y is a linear combination of indicator functions 1_{B_k}, it is sufficient to prove that

$$\mathcal{E}(1_{B_k}X \mid \mathcal{P}) = 1_{B_k}\mathcal{E}(X \mid \mathcal{P})$$

for all $k = 1, \ldots, n$. But this will hold if and only if it holds when multiplied by *each* indicator random variable 1_{B_j}; that is, if and only if

$$\mathcal{E}(1_{B_k}X \mid \mathcal{P})1_{B_j} = 1_{B_k}\mathcal{E}(X \mid \mathcal{P})1_{B_j}$$

for all $j = 1, \ldots, n$. But this is equivalent to

$$\mathcal{E}(1_{B_k}X \mid B_j)1_{B_j} = 1_{B_k}\mathcal{E}(X \mid B_j)1_{B_j}$$

or, after multiplying by $\mathbb{P}(B_j)$,

$$\mathcal{E}(1_{B_k}1_{B_j}X)1_{B_j} = 1_{B_j}1_{B_k}\mathcal{E}(X1_{B_j})$$

However, both sides of this are 0 for $k \neq j$ and both sides are also equal when $j = k$. This proves part 2).

To prove 3), since $\mathcal{E}(X \mid \mathcal{P})$ is constant on the blocks of $\mathcal{P}$, it is also constant on the blocks of the finer partition Q; that is, $\mathcal{E}(X \mid \mathcal{P})$ is Q-measurable. Hence, part 2) implies that

$$\mathcal{E}(\mathcal{E}(X \mid \mathcal{P}) \mid Q) = \mathcal{E}(X \mid \mathcal{P})\mathcal{E}(1_\Omega \mid Q) = \mathcal{E}(X \mid \mathcal{P})$$

To prove the other equality

$$\mathcal{E}(\mathcal{E}(X \mid Q) \mid \mathcal{P}) = \mathcal{E}(X \mid \mathcal{P})$$

it is sufficient to show that for each $B_i \in \mathcal{P}$,

$$\mathcal{E}(\mathcal{E}(X \mid Q) \mid B_i) = \mathcal{E}(X \mid B_i)$$

or multiplying by $\mathbb{P}(B_i)$,

$$\mathcal{E}(\mathcal{E}(X \mid Q)1_{B_i}) = \mathcal{E}(X1_{B_i})$$

But if $Q = \{C_1, \ldots, C_m\}$ then

$$\mathcal{E}(\mathcal{E}(X \mid \mathcal{Q})1_{B_i}) = \mathcal{E}\left(\sum_{k=1}^{m} \mathcal{E}(X \mid C_k)1_{C_k}1_{B_i}\right)$$

$$= \mathcal{E}\left(\sum_{C_k \subseteq B_i} \mathcal{E}(X \mid C_k)1_{C_k}\right)$$

$$= \sum_{C_k \subseteq B_i} \mathcal{E}(X \mid C_k)\mathcal{E}(1_{C_k})$$

$$= \sum_{C_k \subseteq B_i} \mathcal{E}(X1_{C_k})$$

$$= \mathcal{E}(X1_{B_i})$$

as desired.

To prove 4), if $\mathcal{P}_X = \{C_1, \ldots, C_m\}$ then X is a linear combination of indicator functions 1_{C_j} and so we need only prove that

$$\mathcal{E}(1_{C_j} \mid \mathcal{P}) = \mathcal{E}(1_{C_j})$$

for all $j = 1, \ldots, m$. But the independence of $\mathcal{P}$ and $\mathcal{P}_X$ imply that

$$\mathcal{E}(1_{C_j}1_{B_i}) = \mathcal{E}(1_{C_j})\mathcal{E}(1_{B_i})$$

and so

$$\mathcal{E}(1_{C_j} \mid B_i) = \frac{\mathcal{E}(1_{C_j}1_{B_i})}{\mathbb{P}(B_i)} = \frac{\mathcal{E}(1_{C_j})\mathcal{E}(1_{B_i})}{\mathcal{E}(1_{B_i})} = \mathcal{E}(1_{C_j})$$

whence

$$\mathcal{E}(1_{C_j} \mid \mathcal{P}) = \sum_{i=1}^{m}\mathcal{E}(1_{C_j} \mid B_i)1_{B_i} = \sum_{i=1}^{m}\mathcal{E}(1_{C_j})1_{B_i} = \mathcal{E}(1_{C_j})\sum_{i=1}^{m}1_{B_i} = \mathcal{E}(1_{C_j})$$

as desired. $\square$

Exercises

1. A pair of fair dice is rolled. Find the probability of getting a sum that is even.
2. Three fair dice are rolled. Find the probability of getting exactly one 6.
3. A basket contains 5 red balls, 3 black balls, and 4 white balls. A ball is chosen at random from the basket.
    a) Find the probability of choosing a red ball.
    b) Find the probability of choosing a white ball or a red ball.
    c) Find the probability of choosing a ball that is not red.
4. A certain true-and-false test contains 10 questions. A student guesses randomly at each question.
    a) What is the probability that he will get all 10 questions correct?

    b)   What is the probability that he will get at least 9 questions correct?

    c)   What is the probability that he will get at least 8 questions correct?

5.   A die has six sides, but two sides have only 1 dot. The other four sides have $2, 3, 4$ and 5 dots, respectively. Assume that each side is equally likely to occur.

    a)   What is the probability of getting a 1?

    b)   What is the probability of getting a 2?

    c)   What is the probability of getting an even number?

    d)   What is the probability of getting a number less than 3?

6.   Four fair coins are tossed. Find the probability of getting exactly 2 heads.

7.   Four fair coins are tossed. Find the probability of getting at least 2 heads.

8.   A fair die is rolled and a card is chosen at random. What is the probability that the number on the die matches the number on the card? (An ace is counted as a one.)

9.   Studies of the weather in a certain city over the last several decades have shown that, for the month of March, the probability of having a certain amount of sun/smog is as follows:

$\mathbb{P}$(full sun/no smog) $= 0.07$, $\mathbb{P}$(full sun/light smog) $= 0.09$,

$\mathbb{P}$(full sun/heavy smog) $= 0.12$, $\mathbb{P}$(haze/no smog) $= 0.09$,

$\mathbb{P}$(haze/light smog) $= 0.07$, $\mathbb{P}$(haze/heavy smog) $= 0.11$,

$\mathbb{P}$(no sun/no smog) $= 0.16$, $\mathbb{P}$(no sun/light smog) $= 0.12$,

$\mathbb{P}$(no sun/heavy smog) $= 0.17$

What is the probability of having a fully sunny day? What is the probability of having at day with some sun? What is the probability of having a day with no or light smog?

10. Show that if X is a random variable on Ω, then

    a) $\{X \in A \cup B\} = \{X \in A\} \cup \{X \in B\}$

    b) $\{X \in A \cap B\} = \{X \in A\} \cap \{X \in B\}$

    c) $\{X \in A \setminus B\} = \{X \in A\} \setminus \{X \in B\}$

11. Let X be a random variable on $(\Omega, \mathbb{P})$. Show that X can be written as a linear combination of indicator functions of the blocks of the partition generated by X.

12. a)   Consider a stock whose current price is 50 and whose price at some fixed time T in the future may be one of the following values: $48, 49, 50, 51$. Suppose we estimate that the probabilities of these stock prices are

$$\mathbb{P}(48) = 0.2$$
$$\mathbb{P}(49) = 0.4$$
$$\mathbb{P}(50) = 0.3$$
$$\mathbb{P}(51) = 0.1$$

If we purchase one share of the stock now, what is the expected return at time T? What is the expected profit?

b) Consider a derivative of the stock in part a) whose return D is a function of the stock price, say

$$D(48) = 2$$
$$D(49) = -1$$
$$D(50) = 0$$
$$D(51) = 3$$

Thus, the return D is a random variable on Ω. What is the expected return of the derivative?

13. Suppose that you roll a fair die once. If the number on the top face of the die is even, you win that amount, in dollars. If it is odd, you lose that amount. What is the expected value of this game? Would you play?

14. For a cost of $1, you can roll a single fair die. If the outcome is odd, you win $2. Would you play? Why?

15. Suppose you draw a card from a deck of cards. You win the amount showing on the card if it is not a face card, and lose $10 if it is a face card. What is your expected value? Would you play this game?

16. An American roulette wheel has 18 red numbers, 18 black numbers and two green numbers. If you bet on red, you win an amount equal to your bet (and get your original bet back) if a red number comes up, but lose your bet otherwise. What is your expected winnings in this game? Is this a fair game?

17. Consider the dart board shown below.

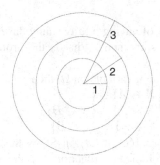

A single dart costs $1.50. You are paid $3.00 for hitting the center, $2.00 for hitting the middle ring and $1.00 for hitting the outer ring. What is the expected value of your winnings? Would you play this game?

18. Prove that $\operatorname{Var}(X) = \mathcal{E}(X)^2 - \mu^2$ where $\mu = \mathcal{E}(X)$.

19. Prove that for any real number a

$$\operatorname{Var}(aX) = a^2 \operatorname{Var}(X)$$

20. Prove that if X and Y are independent random variables then

$$\operatorname{Var}(X + Y) = \operatorname{Var}(X) + \operatorname{Var}(Y)$$

21. Let X be a binomial random variable with distribution $b(k; n, p)$. Show that $\mathrm{Var}(X) = np(1 - p)$. *Hint*: Use the fact that $\mathrm{Var}(X) = \mathcal{E}(X^2) - \mathcal{E}(X)^2$.

22. Show that if X, Y and Z are independent random variables then so are XY and Z.

23. Let X and Y be independent random variables on $(\Omega, \mathbb{P})$. Let f and g be functions from $\mathbb{R}$ to $\mathbb{R}$. Then prove that $f(X)$ and $g(Y)$ are independent.

24. If X is a standard Bernoulli random variable with probabilities p and $q = 1 - p$, show that

$$\mathbb{P}\left(X = \frac{q}{\sqrt{pq}}\right) = p \quad \text{and} \quad \mathbb{P}\left(X = \frac{-p}{\sqrt{pq}}\right) = q$$

25. Prove that the random variables $X_1, \ldots, X_n$ on Ω are independent if and only if the partitions $\mathcal{P}_{X_1}, \ldots, \mathcal{P}_{X_n}$ are independent collections of events.

26. Prove that the relation of refinement is a partial order on $\mathrm{Part}(\Omega)$.

27. Prove that if X is a *nonnegative* random variable; that is, $X(\omega) \geq 0$ for all $\omega \in \Omega$ then $\mathcal{E}(X \mid \mathcal{P}) \geq 0$.

28. A certain operation results in complete recovery 60% of the time, partial recovery 30% of the time and death 10% of the time. What is the probability of complete recovery, given that a patient survives the operation?

29. Imagine the following experiment. You have an unfair coin, whose probabilities are

$$\mathbb{P}(\text{heads}) = \frac{2}{3}, \quad \mathbb{P}(\text{tails}) = \frac{1}{3}$$

You also have two urns containing colored balls, where
1) urn 1 has 3 blue balls and 5 red balls
2) urn 2 has 7 blue balls and 6 red balls
First you toss the coin. If the coin comes up heads, you draw a ball at random from urn 1. If the coin comes up tails, you draw a ball at random from urn 2. What is the probability that the ball drawn is blue? *Hint:* Use the Theorem on Total Probability.

30. Let Ω be a sample space and let $E_1, \ldots, E_n$ form a partition of Ω with $\mathbb{P}(E_k) \neq 0$ for all k. Show that for any event A in Ω,

$$\mathbb{P}(A) = \sum_{k=l}^{n} \mathbb{P}(A \mid E_k)\mathbb{P}(E_k)$$

31. Let $\mathcal{P} = \{B_1, \ldots, B_k\}$ be a partition of $(\Omega, \mathbb{P})$ with $\mathbb{P}(B_i) > 0$ for all i. Prove **Bayes' formula**, which states that for any event A in Ω with $\mathbb{P}(A) > 0$, we have

$$\mathbb{P}(B_j \mid A) = \frac{\mathbb{P}(A \mid B_j)\mathbb{P}(B_j)}{\sum\limits_{i=1}^{k} \mathbb{P}(A \mid B_i)\mathbb{P}(B_i)}$$

32. Prove that if $E_1, \ldots, E_n$ are events with $\mathbb{P}(E_1 \cap \cdots \cap E_n) > 0$ then

$$\mathbb{P}(E_1 \cap \cdots \cap E_n)$$
$$= \mathbb{P}(E_1)\mathbb{P}(E_2 \mid E_1)\mathbb{P}(E_3 \mid E_1 \cap E_2) \cdots \mathbb{P}(E_n \mid E_1 \cap \cdots \cap E_{n-1})$$

33. Prove that if

$$B_0 \supseteq B_1 \supseteq \cdots \supseteq B_n$$

is a nested sequence of events with positive probability in a sample space Ω and if $\mathbb{P}(B_0) = 1$, then

$$\mathbb{P}(B_n) = \mathbb{P}(B_1 \mid B_0)\mathbb{P}(B_2 \mid B_1) \cdots \mathbb{P}(B_n \mid B_{n-1})$$

34. Suppose that words of length 5 over the binary alphabet $\{0, 1\}$ are sent over a noisy communication line, such as a telephone line. Assume that, because of the noise, the probability that a bit (0 or 1) is received correctly is 0.75. Assume also that the event that one bit is received correctly is independent of the event that another bit is received correctly.
    a)   What is the probability that a string will be received correctly?
    b)   What is the probability that exactly 3 of the 5 bits in a string are received correctly?

35. Let X and Y be independent random variables on $(\Omega, \mathbb{P})$. Suppose that X and Y have the same range $\{a_1, \ldots, a_n\}$ and that

$$\mathbb{P}(X = a_i) = \mathbb{P}(Y = a_i) = p_i$$

Compute $\mathbb{P}(X = Y)$.

36. Prove that if A and B are events in Ω with $\mathbb{P}(A \cap B) > 0$, then

$$\mathcal{E}_{\mathbb{P}(\cdot \mid A)}(X \mid B) = \mathcal{E}(X \mid A \cap B)$$

37. Let $X_1, \ldots, X_n$ be random variables on $(\Omega, \mathbb{P})$ all of which have the same expected value μ and the same range $\{r_1, \ldots, r_m\}$. Let N be a random variable on $(\Omega, \mathbb{P})$ where N takes the values $1, \ldots, n$. Assume also that N is independent of the X_i's. Then we can define a random variable S by

$$S = X_1 + \cdots + X_N$$

where

$$S(\omega) = X_1(\omega) + \cdots + X_{N(\omega)}(\omega)$$

Show that
    a) $\mathcal{E}(S \mid N = k) = \mu k$
    b) $\mathcal{E}(S \mid \mathcal{P}_N) = \mu N$, where $\mathcal{P}_N$ is the partition induced by N
    c) $\mathcal{E}(S) = \mu \mathcal{E}(N)$
Explain in words why part c) makes sense.

38. Consider the following game from the game show *Let's Make a Deal*. A contestant is given a choice of three doors, behind one of which is a new car and behind each of the two others is a goat. The contestant wins whatever is

behind the chosen door, once it is opened. Once the contestant chooses a door, host Monty Hall opens one of the two other doors that does *not* conceal the car and asks the contestant if he wishes to change to the other unopened door. If both unchosen doors conceal goats, Hall chooses between the two doors with equal probability. Show that the probability that the contestant finds the car if he does not change doors is $1/3$ but the probability that the contestant finds the car if he does change doors is $2/3$.

Chapter 4
Stochastic Processes, Filtrations and Martingales

State Trees

The following concept will prove quite useful in picturing discrete-time financial models.

As shown in Figure 4.1, a **state tree** $\mathbb{T}$ consists of a finite collection of **nodes** or **vertices** (singular: **vertex**), shown as small boxes in the figure, together with a finite collection of **edges** connecting certain nodes, shown as line segments between nodes.

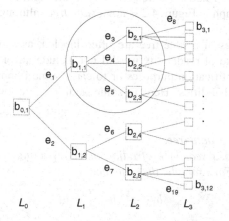

Figure 4.1: A state tree $\mathbb{T}$

The nodes of a state tree are organized in vertical columns called **levels**. The levels of a state tree are often thought of as *times*, where level L_k corresponds to time t_k, the initial level L_0 being the initial (or starting) time t_0 and the final level L_N being the final time t_N. Also, the nodes at time t_k are thought of as the time-t_k **intermediate states** of the tree and the set of all time-t_k nodes is called the **time-t_k state space**. The state space for the final time is called the **final state space**. For example, the time-t_2 state space of the tree in Figure 4.1 is

$$\{b_{2,1}, b_{2,2}, b_{2,3}, b_{2,4}, b_{2,5}\}$$

and the final state space is

$$\{b_{3,1}, \ldots, b_{3,12}\}$$

The edges of a state tree connect nodes of adjacent levels only.

Two nodes in adjacent levels of $\mathbb{T}$ are **adjacent** if there is an edge connecting them. If a node b at level L_k is connected to a node c at level L_{k+1}, then b is called the **parent** of c and c is called a **child** of b. This terminology comes from the fact that trees are often used to show family relationships. For example, referring to Figure 4.1, node $b_{1,1}$ is the parent of $b_{2,2}$ and $b_{2,2}$ is a child of $b_{1,1}$.

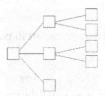

Figure 4.2

In a state tree, the end nodes can appear only in the rightmost level of the tree. For example, the graph in Figure 4.2 is not a *state tree* (although it is a tree).

The nodes and edges of a state tree are often labeled, as shown in Figure 4.1, although for the sake of readability we did not include all of the labels. Labels serve either for identification purposes or to carry some important data, such as stock prices or probabilities. Here is some additional terminology associated with trees.

Definition *Let $\mathbb{T}$ be a state tree.*
1) *A **path** in $\mathbb{T}$ is a sequence of adjacent nodes along with the edges that connect them. For example, the sequence*

$$b_{1,1} \xrightarrow{e_5} b_{2,3} \xrightarrow{e_{14}} b_{3,7}$$

*is a path in $\mathbb{T}$ of **length** 2 from $b_{1,1}$ to $b_{3,7}$. Sometimes, we think of a path as being just the sequence of nodes or just the sequence of edges in the path. If there is a path from node b at level L_k to a node c at a level L_{k+i}, then c is called a **descendent** of b at level L_{k+i}.*
2) *A node b together with the children of b and the edges that connect these children to the parent b form the **child subtree** of b. For example, the*

circled portion of Figure 4.1 is the child subtree of $b_{1,1}$. Note that the child subtree includes the parent node. $\square$

Information Structures

Let $\Omega = \{\omega_1, \ldots, \omega_m\}$ be a finite set. Figure 4.3 shows an example of a state tree in which the time-t_k intermediate states are the blocks of a partition $\mathcal{P}_k$ of Ω and where each partition $\mathcal{P}_k$ is a *refinement* of the previous partition $\mathcal{P}_{k-1}$. This situation is very important in the study of discrete-time financial models and has a special name.

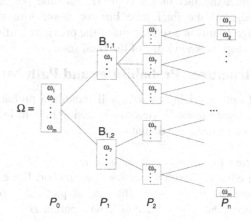

Figure 4.3: A filtration

Definition *A sequence* $\mathbb{F} = (\mathcal{P}_0, \ldots, \mathcal{P}_N)$ *of partitions of a set* $\Omega = \{\omega_1, \ldots, \omega_m\}$ *for which*

$$\mathcal{P}_0 \succ \mathcal{P}_1 \succ \cdots \succ \mathcal{P}_N$$

is called a **filtration**. *A filtration is called an* **information structure** *if* $\mathcal{P}_0$ *is the coarsest possible partition*

$$\mathcal{P}_0 = \{\Omega\}$$

and $\mathcal{P}_N$ *is the finest possible partition*

$$\mathcal{P}_N = \{\{\omega_1\}, \ldots, \{\omega_m\}\}$$

The graph in Figure 4.3 is called a **state tree** *for the filtration.* $\square$

Note that while the final state space

$$\mathcal{P}_N = \{\{\omega_1\}, \ldots, \{\omega_m\}\}$$

in an information structure is different from the underlying set

$$\Omega = \{\omega_1, \ldots, \omega_m\}$$

the difference has no significance for us and so, as is customary, we will generally think of the final state space as the set Ω.

The motivation behind filtrations and information structures is the following. We take the point of view that each element $\omega \in \Omega$ represents a *final state* of some game (or of the economy) and that at each time t_k, we have *partial knowledge* of the final state in that we know which block of $\mathcal{P}_k$ contains the final state.

When $\mathbb{F}$ is an information structure, then at time t_0, we have no knowledge of the final state, other than the fact that it is in Ω. As time passes, we may gain additional knowledge of the final state but we never lose knowledge. This increased knowledge results in a refinement of the previous partition. At the final time t_N, we have *complete* knowledge of the final state.

Information Structures, Probabilities and Path Numbers

Our discussion of option pricing models will include a probability measure $\mathbb{P}$ defined on the final state space Ω. Before discussing how to incorporate such a probability measure, we need two definitions.

Let $\mathbb{T}$ be a state tree for an information structure $\mathbb{F} = (\mathcal{P}_0, \ldots, \mathcal{P}_N)$. Suppose that we label each edge in $\mathbb{T}$ with a *positive* real number. For each intermediate state $B_k \in \mathcal{P}_k$, there are two numbers that are of particular importance to us. First, Figure 4.4 shows the child subtree of $\mathbb{T}$ with parent B_k.

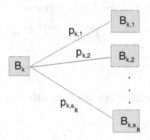

Figure 4.4: The child subtree of B_k

where

$$\text{Children}(B_k) = \{B_{k,1}, \ldots, B_{k,s_k}\}$$

We call the *sum* of the edge labels in the child subtree for B_k,

$$C(B_k) = \sum_{i=1}^{s_k} p_{k,i}$$

the **child subtree number** for B_k.

Second, as shown in Figure 4.5, there is a unique path in the state tree from the root node Ω to B_k.

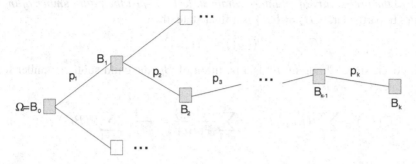

Figure 4.5

We say that the path

$$\Omega = B_0 \to B_1 \to B_2 \to \cdots \to B_k$$

terminates in $\mathcal{P}_k$. The *product* (not the sum) of the edge labels of this path is called the **path number** of B_k. We denote this path number by $H(B_k)$.

Now, there is a simple relationship between path numbers and child subtree numbers. If the edge label of the edge from B_k to B_{k+1} is p, then

$$H(B_{k+1}) = H(B_k)p$$

and so using the notation above,

$$\sum_{i=1}^{s_k} H(B_{k,i}) = \sum_{i=1}^{s_k} H(B_k)p_{k,i} = H(B_k)\sum_{i=1}^{s_k} p_{k,i} = H(B_k)C(B_k) \quad (4.1)$$

In words, the sum of the path numbers of the children of B_k is equal to the path number of B_k times the child subtree number of B_k. Indeed, this is really just the distributive law of arithmetic.

Probability Measures and Edge Labels

Now suppose that $\mathbb{P}$ is a strongly positive probability measure on the state space Ω. Consider a path

$$\Omega = B_0 \to B_1 \to B_2 \to \cdots \to B_k$$

starting at Ω and ending at an intermediate state $B_k \in \mathcal{P}_k$. Because $\mathbb{F}$ is a filtration, the states in this path form a nested sequence of events on Ω with positive probability,

$$\Omega = B_0 \supseteq B_1 \supseteq \cdots \supseteq B_k$$

We asked the reader in an earlier exercise to show that

$$\mathbb{P}(B_k) = \mathbb{P}(B_1 \mid B_0)\mathbb{P}(B_2 \mid B_1)\cdots\mathbb{P}(B_k \mid B_{k-1}) = H(B_k)$$

and so *the probability of an intermediate state B_k is just the path number of that state*. In particular, if $B_k = \{\omega_k\}$ is a final state, then

$$\mathbb{P}(\{\omega\}) = H(\{\omega\})$$

Moreover, since $\text{Children}(B_k)$ is a partition of B_k, the child subtree number for any state $B_k \in \mathcal{P}_k$ is

$$C(B_k) = \sum_{i=1}^{s_k}\mathbb{P}(B_{k,i} \mid B_k) = \sum_{i=1}^{s_k}\frac{\mathbb{P}(B_{k,i})}{\mathbb{P}(B_k)} = \frac{1}{\mathbb{P}(B_k)}\sum_{i=1}^{s_k}\mathbb{P}(B_{k,i}) = 1$$

that is,

$$C(B_k) = 1 \tag{4.2}$$

for all $k = 0, \ldots, N - 1$.

Edge Labels and Probability Measures

Now suppose that instead of starting with a strongly positive probability measure $\mathbb{P}$ on Ω, we simply label each edge of the state tree $\mathbb{T}$ with a positive real number, subject only to the condition that (4.2) holds. Then (4.1) implies that the sum of the path numbers of the children of $B_k \in \mathcal{P}_k$ is equal to the path number $H(B_k)$ of the parent, in symbols,

$$\sum_{i=1}^{s_k}H(B_{k,i}) = H(B_k)$$

From this, we can deduce two important facts.

First, the sum of the path numbers of *all* states in $\mathcal{P}_{k+1}$ (all children) is the same as the sum of the path numbers of all states in $\mathcal{P}_k$ (all parents), in symbols,

$$\sum_{B_{k+1}\in\mathcal{P}_{k+1}}H(B_{k+1}) = \sum_{B_k\in\mathcal{P}_k}H(B_k)$$

for all $k = 1, \ldots, N - 1$. But the sum of the path numbers for the states in the *second* partition $\mathcal{P}_1$ is the child subtree number $C(\Omega)$, which is equal to 1 by assumption. Hence, the sum of the path numbers for all states at any given level is 1. In particular, this applies to the final state space Ω and so

$$\sum_{\omega\in\Omega}H(\{\omega\}) = 1$$

Hence, the path numbers

$$\{H(\{\omega\}) \mid \omega \in \Omega\}$$

form a probability distribution on Ω, with associated probability measure defined by

$$\mathbb{P}(\{\omega\}) = H(\{\omega\})$$

Second, not only is the path number $H(B_k)$ equal to the sum of the path numbers for all of the children of B_k, but equation (4.1) also implies that $H(B_k)$ is equal to the sum of the path numbers of all *grandchildren* of B_k, or indeed of all descendents of B_k at *any* given level of the tree. Thus, since the descendents of the intermediate state B_k at the final level Ω are essentially just the elements of B_k, we have

$$H(B_k) = \sum_{\omega \in B_k} H(\{\omega\}) = \sum_{\omega \in B_k} \mathbb{P}(\{\omega\}) = \mathbb{P}(B_k)$$

and so

$$\mathbb{P}(B_k) = H(B_k)$$

for all states B_k in all partitions $\mathcal{P}_k$.

Finally, it follows that the edge labels are indeed conditional probabilities, since if the edge label from B_k to B_{k+1} is p_k, then

$$p_k = \frac{H(B_{k+1})}{H(B_k)} = \frac{\mathbb{P}(B_{k+1})}{\mathbb{P}(B_k)} = \mathbb{P}(B_{k+1} \mid B_k)$$

Summary

In summary, if we begin with a strongly positive probability measure $\mathbb{P}$ on Ω and label the edges of the state tree with conditional probabilities, then $\mathbb{P}$ is just the path number function; that is,

$$\mathbb{P}(B_k) = H(B_k)$$

for any state $B_k \in \mathcal{P}_k$. On the other hand, if we begin by labeling the edges of the state tree with positive numbers subject only to the condition that all child subtree numbers are equal to 1, then the path number function H defines a strongly positive probability measure $\mathbb{P}$ on Ω and the edge labels are the conditional probabilities with respect to $\mathbb{P}$. Thus, the two approaches lead to the same end result.

Theorem 4.1 *Let* $\Omega = \{\omega_1, \ldots, \omega_m\}$ *be a finite set with information structure*

$$\mathbb{F} = (\mathcal{P}_0, \ldots, \mathcal{P}_N)$$

Suppose that we label the edges of the state tree of $\mathbb{F}$ *with positive real numbers in such a way that each child subtree number is equal to 1; that is,*

$$C(B_k) = 1$$

for all $B_k \in \mathcal{P}_k$ *and all* $k = 0, \ldots, N - 1$. *Then the path number function* H

defines a strongly positive probability distribution on Ω, *with associated probability measure*

$$\mathbb{P}(\{\omega\}) = H(\{\omega\})$$

and more generally,

$$\mathbb{P}(B_k) = H(B_k)$$

for all states $B_k \in \mathcal{P}_k$ *and all* $k = 0, \ldots, N$. *Moreover, the edge label* p_k *from* B_k *to* $B_{k,i}$ *is the conditional probability*

$$p_k = \mathbb{P}(B_{k,i} \mid B_k) \qquad\qquad \square$$

Information Structures and Stochastic Processes

We have seen how to incorporate a probability measure into an information structure. Let us turn now to the issue of incorporating prices and other values into information structures. In particular, we will be interested in the price of a stock as it progresses through a sequence of times

$$t_0 < t_1 < t_2 < \cdots < t_N$$

If the stock price at time t_i is denoted by X_i, then initially these prices are unknown and can be thought of as random variables on the state space. This leads to the following simple concept, which plays a very important role in many areas of applied mathematics, including the mathematics of finance.

Definition *A* (**finite**) **stochastic process** *on a sample space* Ω *is a sequence*

$$\mathbb{X} = (X_0, X_1, \ldots, X_N)$$

of random variables defined on Ω. $\square$

We will have considerable use for the following notion.

Definition *Let*

$$\mathbb{X} = (X_0, X_1, \ldots, X_N)$$

be a stochastic process on Ω. *If* $k \leq m$, *the* **change** *in* $\mathbb{X}$ *from* k *to* m *is the difference*

$$\Delta_{k,m}(\mathbb{X}) = X_m - X_k \qquad\qquad \square$$

In the theory of stochastic process, a change is referred to as an **increment** *of* $\mathbb{X}$, *but we will not use this term.*

Stochastic Processes Adapted to a Filtration

Now consider a filtration

$$\mathbb{F} = (\mathcal{P}_0, \mathcal{P}_1, \ldots, \mathcal{P}_N)$$

on a finite sample space Ω, along with a stochastic process

$$\mathbb{X} = (X_0, X_1, \ldots, X_N)$$

of the same length. What does it mean to say that the value of X_k can be determined at time t_k? At time t_k we know only the *blocks* of the partition $\mathcal{P}_k$ but the random variable X_k is defined on Ω. Hence, the answer is that we can determine the value of X_k if and only if it is constant on each block of $\mathcal{P}_k$; that is, if and only if X_k is $\mathcal{P}_k$-measurable.

For example, suppose that $\Omega = \{1, 2, 3, 4, 5, 6, 7\}$ and that

$$\mathcal{P}_2 = \{B_1 = \{1, 2\}, B_2 = \{3, 4, 5\}, B_3 = \{6, 7\}\}$$

Then we know the value of X_2, given a block of $\mathcal{P}_2$, if and only if X_2 takes the *same* value on 1 and 2, the *same* value on 3, 4 and 5 and the *same* value on 6 and 7. Then, given the block B_i, we can indeed determine the value of X_2.

This brings us to a key definition.

Definition *Let Ω be a finite set, with filtration*

$$\mathbb{F} = (\mathcal{P}_0, \mathcal{P}_1, \ldots, \mathcal{P}_N)$$

A stochastic process

$$\mathbb{X} = (X_0, X_1, \ldots, X_N)$$

on Ω is **adapted** *to the filtration $\mathbb{F}$, or is* **$\mathbb{F}$-adapted** *if X_k is $\mathcal{P}_k$-measurable for all $k = 0, \ldots, N$.* $\square$

Thus, in loose terms, to say that a stochastic process $\mathbb{X}$ is adapted to a filtration is to say that for each k, it is sufficient to know the block of $\mathcal{P}_k$ in order to determine the value of X_k.

We strongly urge the reader to ponder this concept until it seems very clear before continuing. To help put this notion in perspective, we define a related notion that will be important in the next chapter.

Definition *A stochastic process*

$$\mathbb{A} = (A_1, \ldots, A_T)$$

is **predictable** *or* **previewable** *with respect to the filtration*

$$\mathbb{F} = (\mathcal{P}_0, \mathcal{P}_1, \ldots, \mathcal{P}_T)$$

if A_k is $\mathcal{P}_{k-1}$-measurable for all $k = 1, \ldots, T$. □

For a predictable process, we can determine the value of A_k at the *prior time* t_{k-1}. For example, imagine a game where we spin a roulette wheel once at each of the N times $t_1 < t_2 < \cdots t_N$. We can bet on each spin of the wheel. Then we know the outcome X_k of a spin at time t_k but the bet must be made *before* the spin and so we know the amount A_k of the bet on the kth spin at time t_{k-1}. Thus, the *outcomes* are adapted but the *bets* are predictable.

Martingales

Consider the following game between players A and B. A coin is flipped. If the result is heads, player B pays player A one dollar. If the result is tails, player A pays player B one dollar. Let the probability of heads be p. Is this a *fair* game? That is, are both sides willing to play this game?

Of course, both players will play provided that neither player has an advantage and this holds if and only if the expected winnings for either (and therefore both) players is 0. Now, the expected winnings for player A is

$$p - (1 - p) = 2p - 1$$

which is 0 if and only if $p = 1/2$. Thus, as expected, the game is fair if and only if the coin is fair.

In order to generalize this notion of a fair game, let us recast the analysis above in more formal terms. Let X_1 denote the player A winnings after the flip. This is a random variable that takes on two values:

$$X_1(\text{heads}) = 1 \quad \text{and} \quad X_1(\text{tails}) = -1$$

For consistency, let X_0 be the "winnings" for player A before the flip. Thus,

$$X_0(\text{heads}) = 0 \quad \text{and} \quad X_0(\text{tails}) = 0$$

Then the statement that the game is fair can now be written

$$\mathcal{E}(X_1) = X_0$$

In words, the *expected* time-t_1 winnings is equal to the previous time-t_0 winnings. Put another way, the expected *change* in winnings is 0.

Now we can generalize this to describe what it means for a multistage game to be fair. Let $(\Omega, \mathbb{P})$ be a finite probability space and suppose that the stochastic process

$$\mathbb{X} = (X_0, X_1, \ldots, X_N)$$

on Ω is adapted to the filtration

$$\mathbb{F} = (\mathcal{P}_0, \mathcal{P}_1, \ldots, \mathcal{P}_N)$$

on Ω. The reader can think of X_k as the time-t_k winnings in an N-stage game, such as the game of flipping a coin N times.

The mathematical notion that describes the fact that the stochastic process $\mathbb{X}$ is *fair* from one time period to the next is that for each time t_k, the conditional expected value of the time-t_{k+1} winnings X_{k+1} given the current state $\mathcal{P}_k$ of the game is equal to the current winnings X_k; in symbols,

$$\mathcal{E}(X_{k+1} \mid \mathcal{P}_k) = X_k$$

for all $k = 0, \ldots, N - 1$. This leads to the following definition.

Definition *Let* $(\Omega, \mathbb{P})$ *be a finite probability space. Let*

$$\mathbb{F} = (\mathcal{P}_0, \mathcal{P}_1, \ldots, \mathcal{P}_N)$$

be a filtration on Ω and let

$$\mathbb{X} = (X_0, X_1, \ldots, X_N)$$

be a stochastic process adapted to $\mathbb{F}$. *Then* $\mathbb{X}$ *is a* **martingale** *with respect to the triple* $(\Omega, \mathbb{P}, \mathbb{F})$, *if*

$$\mathcal{E}_{\mathbb{P}}(X_{k+1} \mid \mathcal{P}_k) = X_k$$

or equivalently in terms of expected changes,

$$\mathcal{E}_{\mathbb{P}}(\Delta_{k,k+1}(\mathbb{X}) \mid \mathcal{P}_k) = 0$$

for all $k = 0, \ldots, N - 1$. *This expresses the idea that* $\mathbb{X}$ *is "fair" over every one-period time interval* $[t_{k+1}, t_k]$. $\square$

The martingale condition

$$\mathcal{E}_{\mathbb{P}}(\Delta_{k,k+1}(\mathbb{X}) \mid \mathcal{P}_k) = 0$$

is equivalent to the statement that

$$\mathcal{E}_{\mathbb{P}}(\Delta_{k,k+1}(\mathbb{X}) \mid B_k) = 0 \qquad (4.3)$$

for all $B_k \in \mathcal{P}_k$, which says that the expected change in $\mathbb{X}$ on each intermediate state B_k is 0. Since X_k is constant on B_k, if we denote this constant by

$$X_k(B_k)$$

then the martingale condition can also be written

$$\mathcal{E}(X_{k+1} \mid B_k) = X_k(B_k) \qquad (4.4)$$

This says that the expected value of X_{k+1} under the conditional probability distribution $\mathbb{P}(\cdot \mid B_k)$ is just the value X_k on B_k. (Strictly speaking if X_k takes the constant value c on B_k, then $X_k(B_k)$ is equal to $\{c\}$, not c. However, we will allow this minor abuse for readability sake.)

This expresses that fact that $\mathbb{X}$ is "fair" at each intermediate state B_k in $\mathcal{P}_k$. Let us look a bit more closely at this condition. The state $B_k \in \mathcal{P}_k$ is partitioned into its children at time t_{k+1}. The random variable X_k is constant on B_k and the random variable X_{k+1} is constant on each child of B_k, but may take different constant values on different children. So, we have fairness at B_k when the *average* (expected value) of the values $X_{k+1}(B_{k,i})$ is equal to $X_k(B_k)$, under the assumption that B_k has occured.

We will refer to either of the equivalent equations (4.3) and (4.4) as the **local martingale condition** at $B_k \in \mathcal{P}_k$.

Characterizing Martingales

We can now characterize the martingale condition in a variety of useful ways. For example, the martingale condition can be written

$$\mathcal{E}(X_k \mid \mathcal{P}_{k-1}) = X_{k-1}$$

for all $k \geq 1$. Taking conditional expectation with respect to $\mathcal{P}_{k-2}$ and using the tower property, we get

$$\mathcal{E}(X_k \mid \mathcal{P}_{k-2}) = \mathcal{E}(\mathcal{E}(X_k \mid \mathcal{P}_{k-1}) \mid \mathcal{P}_{k-2}) = \mathcal{E}(X_{k-1} \mid \mathcal{P}_{k-2}) = X_{k-2}$$

We can repeat this argument (or use induction) to deduce that

$$\mathcal{E}(X_k \mid \mathcal{P}_j) = X_j \tag{4.5}$$

for any $j < k$. Thus, the martingale condition is equivalent to (4.5) holding for all $j < k$.

Of course, if (4.5) holds for all k, then in particular

$$\mathcal{E}(X_N \mid \mathcal{P}_j) = X_j$$

for all j. Actually, this alone is enough to imply that $\mathbb{X}$ is a martingale, for if this holds for all j, then taking the conditional expectation with respect to $\mathcal{P}_{j-1}$ gives

$$X_{j-1} = \mathcal{E}(X_N \mid \mathcal{P}_{j-1}) = \mathcal{E}(\mathcal{E}(X_N \mid \mathcal{P}_j) \mid \mathcal{P}_{j-1}) = \mathcal{E}(X_j \mid \mathcal{P}_{j-1})$$

and so $\mathbb{X}$ is a martingale.

Theorem 4.2 *Let $(\Omega, \mathbb{P})$ be a finite probability space. Let*

$$\mathbb{F} = (\mathcal{P}_0, \mathcal{P}_1, \dots, \mathcal{P}_N)$$

be a filtration on Ω and let

$$\mathbb{X} = (X_0, X_1, \ldots, X_N)$$

be a stochastic process adapted to $\mathbb{F}$. The following are equivalent:
1) $\mathbb{X}$ is a martingale,

$$\mathcal{E}(X_{k+1} \mid \mathcal{P}_k) = X_k$$

or in terms of change,

$$\mathcal{E}(\Delta_{k,k+1}(\mathbb{X}) \mid \mathcal{P}_k) = 0$$

for all $k = 0, \ldots, N - 1$; that is, $\mathbb{X}$ is "fair" over any one-period time interval $[t_{k+1}, t_k]$.
2) $\mathbb{X}$ is "fair" over any time interval $[t_k, t_{k+i}]$; that is,

$$\mathcal{E}(X_{k+i} \mid \mathcal{P}_k) = X_k \tag{4.6}$$

or in terms of change,

$$\mathcal{E}(\Delta_{k,k+i}(\mathbb{X}) \mid \mathcal{P}_k) = 0$$

for all $i \geq 0$ and $k \geq 0$ for which $k + i \leq N$.
3) $\mathbb{X}$ is "fair" over any time interval of the form $[t_k, t_N]$; that is,

$$\mathcal{E}(X_N \mid \mathcal{P}_k) = X_k$$

or in terms of change,

$$\mathcal{E}(\Delta_{k,N}(\mathbb{X}) \mid \mathcal{P}_k) = 0$$

for all $k = 0, \ldots, N - 1$.
4) $\mathbb{X}$ is fair at every intermediate state $B_k \in \mathcal{P}_k$; that is, the local martingale condition

$$\mathcal{E}(X_{k+1} \mid B_k) = X_k(B_k)$$

holds for all $k = 0, \ldots, N - 1$ and for all states $B_k \in \mathcal{P}_k$.
Moreover, if $\mathbb{X}$ is a martingale, then

$$\mathcal{E}(X_k) = \mathcal{E}(X_0) = X_0(\Omega) \tag{4.7}$$

or in terms of change,

$$\mathcal{E}(\Delta_{0,k}(\mathbb{X})) = 0$$

for all $0 \leq k \leq N$.
Proof. Everything has been proven except the final statement. To see that (4.7) holds, note that (4.6) and the fact that X_0 is constant on Ω give for $k = 0$,

$$\mathcal{E}(X_i)1_\Omega = \mathcal{E}(X_i \mid \mathcal{P}_0) = X_0 \qquad \square$$

As we will see in the next chapter, the fact that X_k can be computed from X_N by taking conditional expectation; that is, the fact that

$$X_k = \mathcal{E}(X_N \mid \mathcal{P}_k)$$

is very important to the matter of option pricing. Indeed, if we think of X_N as the *known* final value of a financial instrument, such as the final value $X_N = (S_T - K)^+$ of a call option, then we will see that any prior value of the option can be computed from this final value in a fair (no-arbitrage) market.

An Example

Let us generalize the coin-tossing example described earlier. If $\omega = a_1 \cdots a_n$ is a string of length n, we use $[\omega]_k$ to denote the prefix $a_1 \cdots a_k$ of ω of length k; that is,

$$[a_1 \cdots a_n]_k = a_1 \cdots a_k$$

Now consider the following game. During each time interval (t_{k-1}, t_k), where

$$t_0 < t_1 < \cdots < t_N$$

a coin is tossed and a record is kept of the results. Thus, after k tosses, we have recorded a string of H's and T's of length k. Let

$$\Omega_k = \{H, T\}^k$$

be the set of all sequences of H's and T's of length k, for $k = 1, \ldots, N$. We denote Ω_N simply by Ω.

At first glance, one might be tempted to define the time-t_k intermediate states of the game as the sequences $\delta \in \Omega_k$. However, in order to form an information structure, we need to think in terms of partitions of the set Ω. Accordingly, we will define the time-t_k state associated with each string $\delta \in \Omega_k$ as the subset

$$\mathcal{F}_k(\delta) = \{\omega \in \Omega \mid [\omega]_k = \delta\}$$

of all strings in Ω with *prefix* δ. Also, we think of the final states of the game as the elements $\omega \in \Omega$, rather than the one-element subsets $\{\omega\}$.

For example, as shown in Figure 4.6, the two time-t_1 states are

$$\mathcal{F}_1(H) = \{HHH, HHT, HTH, HTT\}$$

and

$$\mathcal{F}_1(T) = \{THH, THT, TTH, TTT\}$$

At time t_0, there is only one initial state, which is the entire set Ω.

At any time t_k, the 2^k subsets $\mathcal{F}_k(\delta)$ form a partition

$$\mathcal{P}_k = \{\mathcal{F}_k(\delta_1), \ldots, \mathcal{F}_k(\delta_{2^k})\}$$

of Ω, where $\delta_1, \ldots, \delta_{2^k}$ are the elements of $\{H, T\}^k$. It is clear that $\mathcal{P}_k$ is a refinement of $\mathcal{P}_{k-1}$ and that

$$\mathbb{F} = (\mathcal{P}_0, \mathcal{P}_1, \ldots, \mathcal{P}_N)$$

is an information structure on Ω. Figure 4.6 shows the state tree for the game when $N = 3$. We have labeled each node with both the prefix $\delta \in \Omega_k$ and the state $\mathcal{F}_k(\delta) \in \mathcal{P}_k$.

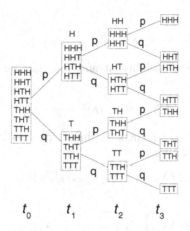

Figure 4.6: State tree for the game

Now let us suppose that for each heads, a player wins 1 dollar and for each tails the player loses 1 dollar. Let X_k be the random variable on Ω that denotes the player's winnings at time t_k. In particular, for any $\omega \in \Omega$, let

$$X_k(\omega) = N_H([\omega]_k) - N_T([\omega]_k)$$

where N_H denotes the number of heads and N_T denotes the number of tails. Thus,

$$\mathbb{X} = (X_0, X_1, \ldots, X_N)$$

is a stochastic process on Ω, where $X_0 = 0$. Moreover, each X_k is $\mathcal{P}_k$-measurable, since every element $\omega \in \mathcal{F}_k(\delta)$ has the same prefix δ and so X_k is constant on $\mathcal{F}_k(\delta)$. Thus, $\mathbb{X}$ is adapted to the filtration $\mathbb{F}$.

Let us now assume that the probability of getting heads is p and that the coin tosses are independent. Then we can define a probability measure on Ω by

$$\mathbb{P}(\{\omega\}) = p^{N_H(\omega)} q^{N_T(\omega)}$$

for $\omega \in \Omega$, where $q = 1 - p$. Each state $\mathcal{F}_k(\delta) \in \mathcal{P}_k$ has exactly two children

$$\mathcal{F}_{k+1}(\delta H) \quad \text{and} \quad \mathcal{F}_{k+1}(\delta T)$$

and the conditional probabilities are

$$\mathbb{P}(\mathcal{F}_{k+1}(\delta H) \mid \mathcal{F}_k(\delta)) = p \quad \text{and} \quad \mathbb{P}(\mathcal{F}_{k+1}(\delta T) \mid \mathcal{F}_k(\delta)) = q$$

so we label the edges of the state tree using these conditional probabilities as follows:

$$\mathcal{F}_k(\delta) \xrightarrow{p} \mathcal{F}_{k+1}(\delta H)$$

and

$$\mathcal{F}_k(\delta) \xrightarrow{q} \mathcal{F}_{k+1}(\delta T)$$

We have seen that the probability of any intermediate state is just its path number and so

$$\mathbb{P}(\mathcal{F}_k(\delta)) = H(\mathcal{F}_k(\delta)) = p^{N_H(\delta)} q^{N_T(\delta)}$$

To determine whether the winnings stochastic process $\mathbb{X}$ is a martingale, we check the local martingale condition

$$\mathcal{E}(X_{k+1} \mid \mathcal{F}_k(\delta)) = X_k(\mathcal{F}_k(\delta))$$

which is equivalent to

$$\frac{\mathcal{E}(X_{k+1} 1_{\mathcal{F}_k(\delta)})}{\mathbb{P}(\mathcal{F}_k(\delta))} = N_H(\delta) - N_T(\delta) \tag{4.8}$$

For convenience, let the winnings associated with the partial result δ be

$$W(\delta) = N_H(\delta) - N_T(\delta)$$

Now, X_{k+1} takes two values on $\mathcal{F}_k(\delta)$: one value on the child $\mathcal{F}_{k+1}(\delta H)$ and one value on the child $\mathcal{F}_{k+1}(\delta T)$. In particular,

$$X_{k+1}(\mathcal{F}_{k+1}(\delta H)) = W(\delta) + 1$$

and

$$X_{k+1}(\mathcal{F}_{k+1}(\delta T)) = W(\delta) - 1$$

Also,

$$\mathbb{P}(\mathcal{F}_{k+1}(\delta H)) = p\mathbb{P}(\mathcal{F}_k(\delta)) \quad \text{and} \quad \mathbb{P}(\mathcal{F}_{k+1}(\delta T)) = q\mathbb{P}(\mathcal{F}_k(\delta))$$

and so

$$\mathcal{E}(X_{k+1} 1_{\mathcal{F}_k(\delta)}) = (W(\delta) + 1)p\mathbb{P}(\mathcal{F}_k(\delta)) + (W(\delta) - 1)q\mathbb{P}(\mathcal{F}_k(\delta))$$
$$= (W(\delta) + p - q)\mathbb{P}(\mathcal{F}_k(\delta))$$

Hence, the left side of the local martingale condition (4.8) is

$$\frac{\mathcal{E}(X_{k+1} 1_{\mathcal{F}_k(\delta)})}{\mathbb{P}(\mathcal{F}_k(\delta))} = N_H(\delta) - N_T(\delta) + p - q$$

and so we have proved that $\mathbb{X}$ is a martingale if and only if $p = q$; that is, if and only if the coin is fair.

Exercises

1. Consider a game with three times $t_0 < t_1 < t_2$. A pair of fair dice, one red and one blue are rolled at time t_0 but you do not see the result. At time t_1, you are told the number on the red die. Draw a state tree with path probabilities.

2. Consider the following game. At time t_0, a fair coin is flipped. At time t_1 a second coin is flipped. At time t_2, a third coin is flipped if and only if the first two coins turned up heads. Can you construct a state tree for the final outcome of the coin flip?

3. Create a state tree with at least three levels. Label each edge with a positive real number in such a way that the path numbers form a probability distribution for each level (after level 0) but that the probability of some parent is not equal to the sum of the probabilities of its children.

4. Prove directly without using Theorem 4.1 that if $\delta \in \Omega_k$, then

$$\mathbb{P}(\mathcal{F}_k(\delta)) = p^{N_H(\delta)} q^{N_T(\delta)}$$

5. Suppose that $\mathbb{X} = (X_0, \ldots, X_N)$ and $\mathbb{Y} = (Y_0, \ldots, Y_N)$ are $\mathbb{F}$-martingales. Show that

$$\mathbb{X} + \mathbb{Y} = (X_0 + Y_0, \ldots, X_N + Y_N)$$

is also an $\mathbb{F}$-martingale.

Exercises on Submartingales and Supermartingales

Let $\mathbb{X} = (X_0, \ldots, X_N)$ be a stochastic process with respect to a filtration $\mathbb{F} = (\mathcal{P}_0, \ldots, \mathcal{P}_N)$. Then $\mathbb{X}$ is a **submartingale** if $\mathbb{X}$ is adapted to $\mathbb{F}$ and

$$\mathcal{E}(X_{k+1} \mid \mathcal{P}_k) \geq X_k$$

Similarly, $\mathbb{X}$ is a **supermartingale** if $\mathbb{X}$ is adapted to $\mathbb{F}$ and

$$\mathcal{E}(X_{k+1} \mid \mathcal{P}_k) \leq X_k$$

6. Show that $\mathbb{X}$ is a supermartingale if and only if $-\mathbb{X}$ is a submartingale. Are submartingales fair games? Whom do they favor? What about supermartingales?

7. (**Doob Decomposition**) Let $(X_0, \ldots, X_N)$ be an $\mathbb{F}$-adapted stochastic process.
   a) Show that there is a unique martingale $(M_0, \ldots, M_N)$ and a unique predictable process $(A_0, \ldots, A_N)$ such that $X_k = M_k + A_k$ and $A_0 = 0$. *Hint*: Set $M_0 = X_0$ and $A_0 = 0$. Then write

$$X_{k+1} - X_k = M_{k+1} - M_k + A_{k+1} - A_k$$

   and take the conditional expectation with respect to $\mathcal{P}_k$. Use the martingale condition to get an expression for $A_{k+1} - A_k$. Then sum from $k = 0$ to $k = n - 1$. Once you have determined $\mathbb{A}$, set $\mathbb{M} = \mathbb{X} - \mathbb{A}$ and show that $\mathbb{A}$ and $\mathbb{M}$ have the required properties.

b)   Show that if (X_k) is a supermartingale then A_k is nonincreasing (that is $A_{k+1} \le A_k$). What if (X_k) is a submartingale?

8.   Let $\mathbb{X} = (X_k \mid k = 0, \ldots, T)$ be a stochastic process.

a)   Prove that for any partition $\mathcal{P}$

$$\max_k \{\mathcal{E}(X_k \mid \mathcal{P})\} \le \mathcal{E}(\max_k \{X_k\} \mid \mathcal{P})$$

b)   Prove that if $\mathbb{X}$ and $\mathbb{Y} = (Y_0, \ldots, Y_N)$ are submartingales then the process defined by

$$Z_k = \max\{X_k, Y_k\}$$

is also a submartingale.

9.   Define the **positive part** of a random variable X by

$$X^+ = \max\{X, 0\}$$

If $\mathbb{X} = (X_0, \ldots, X_N)$ is a martingale, show that $\mathbb{X}^+ = (X_0^+, \ldots, X_N^+)$ is a submartingale.

Chapter 5
Discrete-Time Pricing Models

We now have the necessary mathematical background to discuss *discrete-time* pricing models; that is, pricing models in which all transactions take place at one of a finite number of times.

The **derivative pricing problem** is the problem of determining a fair initial value of any derivative. The difficulty is that the *final* value of the derivative is not known at time $t = 0$, since it generally depends on the final value of the underlying asset, which depends on the state of the economy at the final time. Thus, we will assume that the final value of the underlying is a *known random variable* on the space of all final states of the economy.

Assumptions

We will make the following basic assumptions for the model.

A Unit of Accounting

All prices are given in terms of some **unit of accounting**, also called the **numeraire**. This numeraire may be dollars, eurodollars, pounds sterling, yen and so on. We will generally think in terms of dollars.

Assumption of a Risk-free Asset

We will assume that there is a special asset called the **risk-free asset**. For each time interval $[t_k, t_{k+1}]$, the risk-free asset increases in value by a factor of $e^{r_k(t_k - t_{k-1})}$, where r_k is the **risk-free rate** for that interval. Furthermore, we assume that the risk-free rates are *known in advance*. Practical examples of securities that are generally considered risk-free are U.S. Treasury bonds and U.S. federally insured deposits.

Additional Assumptions

In addition to the previous assumptions, we also assume the following.

Infinitely Divisible Market

The market is **infinitely divisible**, which means that we can speak of, for example, $\sqrt{2}$ or $-\pi$ shares of a stock.

Frictionless Market

All transactions take place immediately and without any external delays. (In reality, transactions do not occur immediately. In fact, many transactions, such as the purchase of stocks and bonds, have a *settlement time* ranging from 1 to 3 days.)

Perfect Market

The market is **perfect**; that is,
- there are no transaction fees or commissions,
- there are no restrictions on short selling,
- the borrowing rate is the same as the lending rate.

Buy-Sell Parity

As an extension of the notion of a perfect market, we assume that any asset's buying price is equal to its selling price; that is, if an asset can be bought for S, then it can also be sold for S. (In reality, most assets have a **bid price** and an **ask price**. To purchase an asset from a broker, the investor must pay the broker's ask price. When selling an asset to a broker, the investor receives the broker's bid price.)

Prices Are Determined under the No-Arbitrage Assumption

As we have discussed at length, we will do all pricing under the assumption that there are no arbitrage opportunities in the market.

The Basic Model

Here are the basic ingredients of the discrete-time model. We will use the symbol $\mathbb{M}$ to denote such a model.

Time

The model $\mathbb{M}$ has $T + 1$ times

$$t_0 < t_1 < \cdots < t_T$$

Note that there are precisely T time *intervals* $[t_{i-1}, t_i]$ for $i = 1, \ldots, T$.

Assets

The model has a finite number of **basic assets**

$$\mathcal{A} = \{\mathfrak{a}_1, \ldots, \mathfrak{a}_n\}$$

The asset $\mathfrak{a}_1$ is assumed to be the risk-free asset; all other asset are **risky**. As mentioned, we denote the risk-free rate for the time interval $[t_k, t_{k+1}]$ by r_k.

States of the Economy

At the final time t_T, we assume that the economy is in one of m possible **final states**, given by the (final) **state space**

$$\Omega = \{\omega_1, \ldots, \omega_m\}$$

In fact, we will think of both the element ω_i and the singleton *set* $\{\omega_i\}$ as a final state, whichever is more convenient at the time.

To model our partial knowledge of the final state of the economy, we use an information structure

$$\mathbb{F} = (\mathcal{P}_0, \mathcal{P}_1, \ldots, \mathcal{P}_T)$$

on the state space Ω, called the **state information structure** for the model $\mathbb{M}$. The partition

$$\mathcal{P}_i = \{B_{i,1}, \ldots, B_{i,m_i}\}$$

of Ω is called the **time-t_i state partition**. For $i < T$, the blocks of $\mathcal{P}_i$ are called the time-t_i **intermediate states**. Figure 5.1 shows the **state tree** for the model. We will denote the state tree for $\mathbb{M}$ by $\mathbb{T}$.

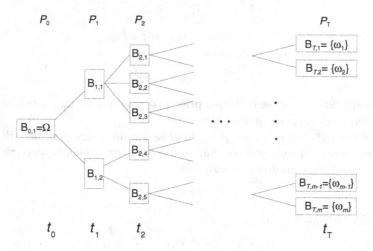

Figure 5.1: The state tree $\mathbb{T}$

Note that since $\mathbb{F}$ is an information structure, we have

$$\mathcal{P}_0 = \{\Omega\} = \{\{\omega_1, \ldots, \omega_n\}\}$$

and

$$\mathcal{P}_T = \{\{\omega_1\}, \ldots, \{\omega_m\}\}$$

Natural Probabilities

We assume the existence of a probability measure on Ω that reflects the likelihood that each final state in Ω will be the actual final state. These are called **natural probabilities**. Natural probabilities are generally either postulated based on economic theory or chosen empirically based on economic data.

Asset Prices

In a discrete-time model, each asset has a price at each time t_i that depends on the state of the economy at that time. This calls for a price random variable $S_{i,j}$ for each time t_i and for each asset a_j. Since we want the price random variables for each asset to form a stochastic process, all price random variables $S_{i,j}$ are defined on the final state space Ω. Moreover, since we know the time price at time t_i, the price $S_{i,j}$ must be $\mathcal{P}_i$-measurable.

Definition
1) *For each time t_i and each asset a_j, the* **price random variable** *$S_{i,j} \colon \Omega \to \mathbb{R}$ is a nonnegative $\mathcal{P}_i$-measurable random variable for which $S_{i,j}(\omega)$ is the time-t_i price of asset a_j under the final state $\omega \in \Omega$.*
2) *For a fixed time t_k, the vector*

$$(S_{k,1}, \ldots, S_{k,n})$$

   of asset prices is called the **price vector** *for time t_k.*
3) *For a fixed asset a_j, the sequence*

$$\mathbb{S}_j = (S_{0,j}, \ldots, S_{T,j})$$

   is a stochastic process, called the **price process** *for asset a_j. It describes the evolution of the price of asset a_j over time.*
4) *For the risk-free asset, the price random variables are constant; that is, they do not depend upon the state of the economy (which is precisely why they are called risk-free). In particular,*

$$S_{0,1} = 1$$

   and for all times $i > 0$,

$$S_{i,1} = e^{r_0(t_1-t_0)} \cdots e^{r_{i-1}(t_i-t_{i-1})}$$

   where r_k is the risk-free rate for the time interval $[t_k, t_{k+1}]$. $\square$

Note that for each i,

$$S_{i,1} = e^{r_{i-1}(t_i-t_{i-1})} S_{i-1,1}$$

Note also that the price processes $\mathbb{S}_j$ are adapted to the filtration $\mathbb{F}$.

Discounted Asset Prices

To compare a payoff of A_i dollars at time t_i with a payoff of A_j dollars at time t_j, we must discount both payoffs to the same time period, which we generally take to be the initial time t_0, *using the risk-free rates.* The time-t_0 **discounted value** of A_i dollars at time t_i is therefore

$$\overline{A}_i = A_i e^{-r_{i-1}(t_i - t_{i-1})} \cdots e^{-r_0(t_1 - t_0)} = \frac{A_i}{S_{i,1}}$$

Thus, the discounted t_0-value of A_i is the time-t_i value of A_i *in the units of the risk-free asset.* We can now say that a payoff of A_i at time t_i is better than a payoff of A_j at time t_j if and only if

$$\overline{A}_i > \overline{A}_j$$

This explains the importance of the following definition.

Definition *For a discrete-time model* $\mathbb{M}$*, the* **discounted asset prices** *are given by*

$$\overline{S}_{i,j} = \frac{S_{i,j}}{S_{i,1}}$$

The value $\overline{S}_{i,j}$ *is also called the (time-t_0)* **present value** *of* $S_{i,j}$. $\square$

Note in particular that

$$\overline{S}_{i,1} = 1$$

and so, as expected, the risk-free asset has a constant unit price in units of itself.

We should also remark that the overbar notation is a bit dangerous, since it does not specify the time period over which the discounting should be computed. Thus, the context must make that clear. For example, $\overline{S}_{i,j}$ means divide by $S_{i,1}$ but $\overline{S}_{i+1,j}$ means divide by $S_{i+1,1}$.

Portfolios and Trading Strategies

In a discrete-time model $\mathbb{M}$ with time periods

$$t_0 < \cdots < t_T$$

an investor is permitted to buy or sell a particular asset of the model at any of the times t_i. Any purchase or sale at time t_i remains in effect until the next time period t_{i+1}; that is, assets are held (as long or short positions) during the entire time interval $[t_i, t_{i+1}]$.

Since it is reasonable to allow these holdings to depend on the state of the economy at the time of acquisition, these holdings are *random variables.*

Moreover, since the quantity of an asset held during a time interval $[t_{i-1}, t_i]$ must be known at the starting time t_{i-1}, these random variables must be $\mathcal{P}_{i-1}$-measurable. Here is the formal definition.

Definition
1) The **asset holding** for asset $\mathfrak{a}_j$ during the time interval $[t_{i-1}, t_i]$ is a $\mathcal{P}_{i-1}$-measurable random variable $\theta_{i,j}$, where $\theta_{i,j}(\omega)$ is the quantity of asset $\mathfrak{a}_j$ held during this time interval under state $\omega \in \Omega$. A positive value of $\theta_{i,j}$ indicates a long position and a negative value of $\theta_{i,j}$ indicates a short position.
2) The **asset holding process** for asset $\mathfrak{a}_j$ is the stochastic process

$$\Phi_j = (\theta_{1,j}, \ldots, \theta_{T,j})$$

3) A **portfolio** for the time interval $[t_{i-1}, t_i]$ is a random vector

$$\Theta_i = (\theta_{i,1}, \ldots, \theta_{i,n})$$

on Ω where $\theta_{i,j}$ is the asset quantity for asset $\mathfrak{a}_j$ over this time interval. $\square$

Rolling Assets

If an asset $\mathfrak{a}_j$ is sold at a given time t_k and the proceeds are used to immediately repurchase the same asset, we say that the asset is **rolled over** at time t_k. (We will explain the reason for this seemingly purposeless action in a moment.) Thus, the quantity of the asset does not change by rolling the asset over and no money is required, because we are assuming that there are no commissions or fees on sales or purchases.

If one or more assets $\mathfrak{a}_{j_1}, \ldots, \mathfrak{a}_{j_p}$ are sold at time t_k and the proceeds are immediately used to acquire an asset $\mathfrak{a}_k$, we say that the assets $\mathfrak{a}_{j_1}, \ldots, \mathfrak{a}_{j_p}$ are **rolled into** asset $\mathfrak{a}_k$.

Portfolio Rebalancing

The use of portfolios in a discrete-time model is a dynamic process that proceeds as follows. At the initial time t_0, the investor acquires the first portfolio

$$\Theta_1 = (\theta_{1,1}, \ldots, \theta_{1,n})$$

which is held through the time interval $[t_0, t_1]$. Note that the random variables $\theta_{1,j}$ are $\mathcal{P}_0$-measurable; that is, constant.

At time t_1, the investor *must* liquidate the portfolio Θ_1 and acquire a new portfolio Θ_2, although there is nothing to prevent the investor from simply rolling over each asset in the portfolio, in symbols, $\Theta_2 = \Theta_1$. Even in this case, however, for reasons of consistency it is simpler to think in terms of complete liquidation followed by acquisition. This does no harm since the model is assumed to be commission-free.

In general, at time t_i the portfolio Θ_i is liquidated and the portfolio Θ_{i+1} is acquired. This process is referred to as **portfolio rebalancing**. The sequence of portfolios obtained through portfolio rebalancing has a name.

Definition *A* **trading strategy** *for a model* $\mathbb{M}$ *is a sequence of portfolios*

$$\Phi = (\Theta_1, \ldots, \Theta_T)$$

where $\Theta_i = (\theta_{i,1}, \ldots, \theta_{i,n})$ *is a portfolio for the time interval* $[t_{i-1}, t_i]$. $\square$

Models are Two-Dimensional

As shown in Figures 5.2 and 5.3, a discrete-time pricing model with at least two assets can be thought of as a two-dimensional structure, with time on the horizontal axis and assets on the vertical axis.

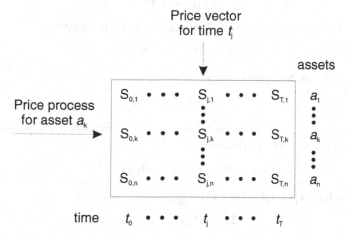

Figure 5.2: Price processes and price vectors

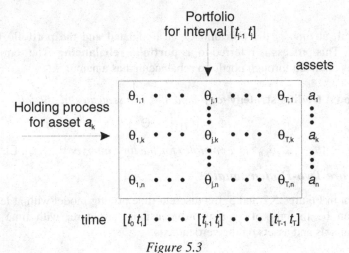

Figure 5.3

For each time t_j, we have two asset-related vectors: a price vector

$$S_j = (S_{j,1}, \ldots, S_{j,n})$$

and a portfolio (or holdings vector)

$$\Theta_j = (\theta_{j,1}, \ldots, \theta_{j,n})$$

for the interval $[t_{j-1}, t_j]$. For each asset a_k, we have two stochastic processes: an $\mathbb{F}$-predictable asset holding process

$$\Phi_k = (\theta_{1,k}, \ldots, \theta_{T,k})$$

and an $\mathbb{F}$-adapted price process

$$\mathbb{S}_k = (S_{0,k}, \ldots, S_{T,k})$$

The Valuation of Portfolios

Since a portfolio Θ_i exists only during the time interval $[t_{i-1}, t_i]$, it makes sense to assign a value to Θ_i only at the **acquisition time** t_{i-1} and the **liquidation time** t_i. The **acquisition value** or **acquisition price** of the portfolio Θ_i is defined by the inner product

$$\mathcal{V}_{i-1}(\Theta_i) = \langle \Theta_i, S_{i-1} \rangle = \sum_{j=1}^{n} \theta_{i,j} S_{i-1,j}$$

and the **liquidation value** or **liquidation price** of Θ_i is defined by

$$\mathcal{V}_i(\Theta_i) = \langle \Theta_i, S_i \rangle = \sum_{j=1}^{n} \theta_{i,j} S_{i,j}$$

We can also discount the portfolio values

$$\overline{V}_{i-1}(\Theta_i) = \langle \Theta_i, \overline{S}_{i-1} \rangle = \sum_{j=1}^{n} \theta_{i,j} \overline{S}_{i-1,j} = \frac{1}{S_{i-1,1}} V_{i-1}(\Theta_i)$$

and

$$\overline{V}_i(\Theta_i) = \langle \Theta_i, \overline{S}_i \rangle = \sum_{j=1}^{n} \theta_{i,j} \overline{S}_{i,j} = \frac{1}{S_{i,1}} V_i(\Theta_i)$$

As with individual asset prices, it makes no sense to compare *nondiscounted* portfolio prices from different time periods; only *discounted* prices can be compared meaningfully.

Note that the discounted value of a quantity $\theta_{k,1}$ of the *risk-free asset* at time t_k is just equal to the quantity $\theta_{k,1}$ itself, since

$$\frac{\theta_{k,1} S_{k,1}}{S_{k,1}} = \theta_{k,1}$$

In short, *for the risk-free asset,*

$$\text{quantity} = \text{discounted value}$$

Thus, if at time t_k, we wish to roll some or all of the assets of a trading strategy into the risk-free asset and if the time-t_k value of these assets is V, after selling the assets, we must buy $\overline{V}$ units of the risk-free asset.

Self-Financing Trading Strategies

For a trading strategy $\Phi = (\Theta_1, \ldots, \Theta_T)$, if the acquisition price of Θ_{i+1} is equal to the liquidation price of Θ_i, then no money is taken out or put into the model during the time-t_i rebalancing process. These are the only types of trading strategies that we will consider.

Definition *A trading strategy* $\Phi = (\Theta_1, \ldots, \Theta_T)$ *is* **self-financing** *if for any time* t_i *where* $i \neq 0, T$, *the acquisition price of* Θ_{i+1} *is equal to the liquidation price of* Θ_i; *that is,*

$$\mathcal{V}_i(\Theta_{i+1}) = \mathcal{V}_i(\Theta_i) \qquad \qquad \square$$

Thus, a self-financing trading strategy is initially purchased for the acquisition value $\mathcal{V}_0(\Theta_1)$ of the first portfolio and is liquidated at time t_T, producing a final payoff of $\mathcal{V}_T(\Theta_T)$. No other money is added to or removed from the model during its lifetime.

The set $\mathcal{T}$ of all *self-financing* trading strategies is a vector space under the operations of coordinate-wise addition

$$(\Theta_{1,1}, \ldots, \Theta_{1,T}) + (\Theta_{2,1}, \ldots, \Theta_{2,T}) = (\Theta_{1,1} + \Theta_{2,1}, \ldots, \Theta_{1,T} + \Theta_{2,T})$$

and scalar multiplication

$$a(\Theta_1, \ldots, \Theta_T) = (a\Theta_1, \ldots, a\Theta_T)$$

Proof of this is left to the reader.

We can extend the use of the symbol $\mathcal{V}_i$ to *self-financing* trading strategies by defining the time-t_i value of Φ to be the common value of the liquidation price of Θ_i and the acquisition price of Θ_{i+1}. In symbols

$$\mathcal{V}_i(\Phi) = \mathcal{V}_i(\Theta_i) = \mathcal{V}_i(\Theta_{i+1})$$

Of course, this makes no sense for non self-financing trading strategies. We will refer to $\mathcal{V}_0(\Phi)$ as the **initial cost** of the trading strategy Φ and to $\mathcal{V}_T(\Phi)$ as the **final payoff** of Φ.

Definition *The stochastic process*

$$\mathbb{V} = (\mathcal{V}_0(\Phi), \ldots, \mathcal{V}_T(\Phi))$$

is called the **value process** *for* Φ. *The* **discounted value process** *is*

$$\overline{\mathbb{V}} = (\overline{\mathcal{V}}_0(\Phi), \ldots, \overline{\mathcal{V}}_T(\Phi)) \qquad\qquad \square$$

Note that since the prices are $\mathbb{F}$-adapted and the quantities are $\mathbb{F}$-predictable, it follows that the value process is also $\mathbb{F}$-adapted.

Example 5.1 One of the simplest trading strategies consists of the initial acquisition of a single unit of asset a_j, which is held throughout the model. Thus,

$$\Theta_i = (0, \ldots, 0, 1, 0, \ldots, 0)$$

for all i, where the 1 appears in the jth coordinate. We denote this trading strategy by $\Phi[a_j]$ and call it the **single-asset trading strategy**. For this trading strategy, the value process is just the price process; that is,

$$(\overline{\mathcal{V}}_0(\Phi), \ldots, \overline{\mathcal{V}}_T(\Phi)) = (\overline{S}_{0,j}, \ldots, \overline{S}_{T,j}) = \overline{\mathbb{S}}_j$$

Thus, a price process is a special type of value process.$\square$

Theorem 5.1
1)   If Θ_1 and Θ_2 are portfolios for the time interval $[t_{i-1}, t_i]$, then

$$\mathcal{V}_{i-1}(a\Theta_1 + b\Theta_2) = a\mathcal{V}_{i-1}(\Theta_1) + b\mathcal{V}_{i-1}(\Theta_2)$$
$$\mathcal{V}_i(a\Theta_1 + b\Theta_2) = a\mathcal{V}_i(\Theta_1) + b\mathcal{V}_i(\Theta_2)$$

for all $a, b \in \mathbb{R}$.

2)   If Φ_1 and Φ_2 are self-financing trading strategies, then

$$\mathcal{V}_i(a\Phi_1 + b\Phi_2) = a\mathcal{V}_i(\Phi_1) + b\mathcal{V}_i(\Phi_2)$$

for all $a, b \in \mathbb{R}$.□

Discounted Gains

The change in value of a portfolio over a period of time is called the *gain*. However, we must be careful to properly discount the values to obtain a meaningful measure of gain.

Definition *Let* Φ *be a self-financing trading strategy.*
1)   *For* $j < k$, *the* **discounted gain** $\overline{G}_{j,k}$ *from time* t_j *to time* t_k *is*

$$\overline{G}_{j,k}(\Phi) = \overline{\mathcal{V}}_k(\Phi) - \overline{\mathcal{V}}_j(\Phi) = \frac{1}{S_{k,1}}\mathcal{V}_k(\Phi) - \frac{1}{S_{j,1}}\mathcal{V}_j(\Phi)$$

2)   *For any time* t_k, *the (cumulative)* **discounted gain** $\overline{G}_k$ *is*

$$\overline{G}_k(\Phi) = \overline{\mathcal{V}}_k(\Phi) - \overline{\mathcal{V}}_0(\Phi)$$

We refer to $\overline{G}_T(\Phi)$ *as the* **discounted final gain** *in* Φ.□

Note that the discounting is always done relative to the *beginning* of the model; that is, to time t_0. Since we are concerned only with discounted gains, we will sometimes use the term *gain* to mean discounted gain. It is clear that gains are additive; that is, for $j \le k \le \ell$ we have

$$\overline{G}_{j,\ell}(\Phi) = \overline{G}_{j,k}(\Phi) + \overline{G}_{k,\ell}(\Phi)$$

It is also clear that the risk-free asset contributes naught to the discounted gain, since the discounted value of the risk-free asset is equal to the quantity, which does not change by itself.

The absolute value of a negative gain is sometimes referred to as a **loss**. For example, a gain of -5 dollars is referred to as a loss of 5 dollars.

Example 5.2 The discounted gain in the single-asset trading strategy $\Phi[a_j]$ is the change in the discounted price of the asset; that is,

$$\overline{G}_{k,k+m}(\Phi[a_j]) = \overline{S}_{k+m,j} - \overline{S}_{k,j} = \Delta_{k,k+m}(\overline{\mathbb{S}}_j) \qquad \square$$

Changing the Initial Value of a Trading Strategy

If $\Phi = (\Theta_1, \ldots, \Theta_T)$ is a self-financing trading strategy, we can create a new self-financing trading strategy simply by changing the initial quantity of the risk-free asset and rolling over this additional quantity throughout the model. This has no effect on the discounted gain and so it can be used to produce another trading strategy with the same gain as the original trading strategy but with 0 initial cost.

In symbols, adjusting the risk-free asset by a gives a new trading strategy $\Phi' = (\Theta'_1, \ldots, \Theta'_T)$ defined by

$$\Theta'_k = \Theta_k + (a1_\Omega, 0, \ldots, 0)$$

for all k. The time-t_k value of Φ' is

$$\mathcal{V}_k(\Phi') = \mathcal{V}_k(\Phi) + aS_{k,1}$$

In particular,

$$\mathcal{V}_0(\Phi') = \mathcal{V}_0(\Phi) + a$$

and so if $a = -\mathcal{V}_0(\Phi)$, then the new trading strategy has zero initial value.

Preserving Gains in a Trading Strategy

Given a self-financing trading strategy, there are ways to isolate and *preserve* any gains made over a given period of time and for certain intermediate states by modifying the original trading strategy. For the present discussion, let $\Phi = (\Theta_1, \ldots, \Theta_T)$ be a self-financing trading strategy.

Locking in a Gain

We say that a self-financing trading strategy $\Phi' = (\Theta'_1, \ldots, \Theta'_T)$ **locks in the gain** in Φ up to time t_k if

$$\overline{G}_T(\Phi') = \overline{G}_k(\Phi)$$

To define Φ', we follow the original strategy Φ until time t_k, at which time we simply roll all assets into the risk-free asset and continue to roll over the risk-free asset for the rest of the time. At time t_k, the portfolio is worth $\mathcal{V}_k(\Theta_k)$ and this is equivalent to $\overline{\mathcal{V}}_k(\Theta_k)$ units of the risk-free asset (recall that for the risk-free asset, quantity equals discounted value). Thus, the new trading strategy $\Phi' = (\Theta'_1, \ldots, \Theta'_T)$ is

$$\Theta'_i = \begin{cases} \Theta_i^\bullet & \text{if } 1 \le i \le k \\ (\overline{\mathcal{V}}_k(\Theta_k), 0, \ldots, 0) & \text{if } k+1 \le i \le T \end{cases}$$

for $k = 0, \ldots, T$. Since the risk-free asset does not produce any discounted gain, we have

$$\overline{G}_T(\Phi') = \overline{G}_k(\Phi') + \overline{G}_{k,T}(\Phi') = \overline{G}_k(\Phi') = \overline{G}_k(\Phi)$$

and so Φ' does indeed lock in the gain of Φ up to time t_k. Note that this makes sense for $k = 0$, since in this case we simply lock in the initial value of Φ by acquiring only the risk-free asset at time t_0 and so

$$\Theta_0' = (\overline{\mathcal{V}}_0(\Theta_0), 0, \ldots, 0)$$

It also makes sense for $k = T$, since in this case we do nothing but follow Φ.

Starting Late Using Cash-and-Carry

The opposite of locking in is starting late. To preserve the gain in Φ *starting* at time t_k, we do nothing until time $t_k < t_T$, at which time we cash-and-carry the portfolio Θ_{k+1}. (This is instead of selling Θ_k and buying Θ_{k+1}, which is what we would have done if we had invested in Φ from the beginning.) Then we follow the trading strategy Φ and simply roll over our debt to the end of the model.

If $\Phi' = (\Theta_1', \ldots, \Theta_T')$ is this new trading strategy, then doing nothing before time t_k means that

$$\Theta_i' = 0 \text{ if } 1 \le i \le k$$

At time t_k, the value of Θ_{k+1} is $\mathcal{V}_k(\Theta_{k+1})$ and so we must borrow the quantity $\overline{\mathcal{V}}_k(\Theta_{k+1})$ of the risk-free asset in order to buy Θ_{k+1}. Hence,

$$\Theta_{k+1}' = \Theta_{k+1} - (\overline{\mathcal{V}}_k(\Theta_{k+1}), 0, \ldots, 0)$$

Since we simply roll over the risk-free asset from that point on, we have

$$\Theta_i' = \Theta_i - (\overline{\mathcal{V}}_k(\Theta_{k+1}), 0, \ldots, 0)$$

for all $i \ge k + 1$. Thus,

$$\Theta_i' = \begin{cases} 0 & \text{if } 1 \le i \le k \\ \Theta_i - (\overline{\mathcal{V}}_k(\Theta_{k+1}), 0, \ldots, 0) & \text{if } k + 1 \le i \le T \end{cases}$$

for $k = 0, \ldots, T - 1$. Note that this makes sense for $k = 0$, since in this case we have

$$\Theta_i' = \Theta_i - (\overline{\mathcal{V}}_0(\Theta_1), 0, \ldots, 0)$$

which is a cash-and-carry of the initial portfolio Θ_1. It also makes sense for $k = T$, in which case starting at the end of the model means doing nothing at all!

Starting Late and Then Locking In

Of course, we can start late and also lock in a gain at a later time. Starting late gives the trading strategy

$$\Theta_i' = \begin{cases} 0 & \text{if } 1 \le i \le k \\ \Theta_i - (\overline{\mathcal{V}}_k(\Theta_{k+1}), 0, \ldots, 0) & \text{if } k + 1 \le i \le T \end{cases}$$

Now, at time t_{k+m}, the value of this trading strategy is

$$\mathcal{V}_{k+m}(\Theta_{k+m}') = \mathcal{V}_{k+m}(\Theta_{k+m}) - \overline{\mathcal{V}}_k(\Theta_{k+1})S_{k+m,1}$$

and so the discounted value is

$$\overline{\mathcal{V}}_{k+m}(\Theta'_{k+m}) = \overline{\mathcal{V}}_{k+m}(\Theta_{k+m}) - \overline{\mathcal{V}}_k(\Theta_{k+1}) = \overline{G}_{k,k+m}(\Phi)$$

Hence, locking in at time t_{k+m} produces the trading strategy

$$\Phi^{(k,k+m)} = (\Theta_1^{(k,k+m)}, \ldots, \Theta_T^{(k,k+m)})$$

given by

$$\Theta_i^{(k,k+m)} = \begin{cases} 0 & \text{for } 1 \le i \le k \\ \Theta_i - (\overline{\mathcal{V}}_k(\Theta_{k+1}), 0, \ldots, 0) & \text{for } k+1 \le i \le k+m \\ (\overline{G}_{k,k+m}(\Phi), 0, \ldots, 0) & \text{if } k+m < i \le T \end{cases}$$

The discounted gain of $\Phi^{(k,k+m)}$ is thus

$$\overline{G}_T(\Phi^{(k,k+m)}) = \overline{G}_{k,k+m}(\Phi)$$

Thus, $\Phi^{(k,k+m)}$ **locks in the gain** in Φ from time t_k to time t_{k+m}. We will denote $\Phi^{(k,k+1)}$ simply by $\Phi^{(k)}$.

Restricting the States

Suppose we want to lock in a gain only for certain states of the economy, say the states

$$\mathcal{A} = \{B_{k_1}, \ldots, B_{k_s}\} \subseteq \mathcal{P}_k$$

It is a simple matter to adjust our trading strategy to restrict activity just to these states: We simply multiply by the indicator function 1_B of the union

$$B = B_{k_1} \cup \cdots \cup B_{k_s}$$

to get

$$\Theta_i^{(k,k+m)} 1_B = \begin{cases} 0 & \text{if } 1 \le i \le k \\ \Theta_i 1_B - (\overline{\mathcal{V}}_k(\Theta_{k+1}), 0, \ldots, 0) 1_B & \text{if } k+1 \le i \le k+m \\ (\overline{G}_{k,k+m}(\Phi), 0, \ldots, 0) 1_B & \text{if } k+m < i \le T \end{cases}$$

The discounted gain of $\Phi^{(k,k+m)} 1_B$ is thus

$$\overline{G}_T(\Phi^{(k,k+m)} 1_B) = \overline{G}_{k,k+m}(\Phi) 1_B$$

and so $\Phi^{(k,k+m)} 1_B$ **locks in the gain** in Φ from time t_k to time t_{k+m} under the states in $\mathcal{A}$ only and has 0 gain for all other states.

We can now summarize.

Theorem 5.2 *Let* $\Phi = (\Theta_1, \ldots, \Theta_T)$ *be a self-financing trading strategy, let* $[t_k, t_{k+m}]$ *be a time interval, let*

$$\mathcal{A} = \{B_{k_1}, \ldots, B_{k_s}\}$$

be a collection of time-t_k states in $\mathcal{P}_k$ and let

$$B = B_{k_1} \cup \cdots \cup B_{k_s}$$

Then the self-financing trading strategy $\Phi^{(k,k+m)}1_B$ defined by

$$\Theta_i^{(k,k+m)}1_B = \begin{cases} 0 & \text{if } 1 \le i \le k \\ \Theta_i 1_B - (\overline{\mathcal{V}}_k(\Theta_{k+1}), 0, \ldots, 0)1_B & \text{if } k+1 \le i \le k+m \\ (\overline{G}_{k,k+m}(\Phi), 0, \ldots, 0)1_B & \text{if } k+m < i \le T \end{cases}$$

locks in the discounted gain in Φ over the interval $[t_k, t_{k+m}]$ for the states in $\mathcal{A}$ only; that is

$$\overline{G}_T(\Phi^{(k,k+m)}1_B) = \overline{G}_{k,k+m}(\Phi)1_B \qquad \qquad \square$$

The following special case of starting late and locking in a gain is especially important.

Example 5.3 (Single-asset, single-period, single-state trading strategy) Let $\Phi[a_j]$ be a single-asset trading strategy. If we start this strategy at time t_k and lock in the gain at the next time t_{k+1} for state $B \in \mathcal{P}_k$ only, the resulting trading strategy is

$$\Theta_i = \begin{cases} 0 & \text{if } 1 \le i \le k \\ (-\overline{S}_{k,j}, \ldots, 0, 1, 0, \ldots, 0)1_B & \text{if } i = k+1 \\ (-\overline{S}_{k,j} + \overline{S}_{k+1,j}, 0, \ldots, 0)1_B & \text{if } k+1 < i \le T \end{cases}$$

This trading strategy, which we denote by $\Phi[a_j, t_k, B]$ is the most "atomic" trading strategy possible: Buy and hold one unit of one asset for one time period in one state only. In fact, we will call these the **atomic trading strategies**. The *final* gain in an atomic trading strategy

$$\overline{G}_T(\Phi[a_j, t_k, B]) = \overline{G}_{k,k+1}(\Phi[a_j])1_B = \Delta_{k,k+1}(\overline{S}_j)1_B$$

is the change in price from t_k to t_{k+1} for asset a_j in state B.$\square$

Arbitrage Trading Strategies

It is now time to formally consider the notion of arbitrage in a discrete-time model. Arbitrage is a situation in which there is no possibility of loss but there is a possibility of a *discounted gain*; that is, a gain *relative to the risk-free asset*. Recall that a random variable X is strictly positive, written $X > 0$, if $X(\omega) \ge 0$ for all $\omega \in \Omega$ and $X(\omega) > 0$ for some $\omega \in \Omega$.

Definition *A self-financing trading strategy Φ is an* **arbitrage trading strategy** *or* **arbitrage opportunity** *if*

$$\overline{G}_T(\Phi) > 0 \qquad \qquad \square$$

Some authors require that an arbitrage trading strategy have initial value 0. However, since we have seen that the initial value of a trading strategy can easily be adjusted to produce a new self-financing trading strategy with initial value 0 without changing the discounted gain, this requirement is not really necessary.

Note that according to the definition, a model will have arbitrage if the gain in any *atomic* trading strategy is strictly positive; that is, if

$$\overline{G}_T(\Phi[\mathfrak{a}_j, t_k, B]) > 0$$

for some asset $\mathfrak{a}_j$, some time t_k and some state $B \in \mathcal{P}_k$. In other words, all it takes to have arbitrage is a *single* asset $\mathfrak{a}_j$ that has a positive discounted gain during a *single* time period $[t_k, t_{k+1}]$ for a *single* intermediate state $B \in \mathcal{P}_k$. This is illustrated in Figure 5.4. In particular, assuming that \$10 grows to a value no greater than \$11 under the risk-free rate, the figure shows that the asset price can rise from \$10 to either \$11, \$12 or \$13. This produces an arbitrage opportunity: Do nothing until time t_k. Then, if the state is B, borrow \$10 and buy the asset. Then sell the asset and repay the debt at time t_{k+1}. In all other situations, do nothing.

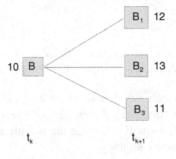

Figure 5.4

On the other hand, a model with at least two non-risk-free assets may have arbitrage even though the *atomic* trading strategies are not arbitrage opportunities. We leave proof of this as an exercise.

However, we can show that an arbitrage trading strategy Φ must exhibit arbitrage on some child subtree of the model; that is, the discounted gain from some parent state to each of its child states must be nonnegative and one such gain must be positive, in symbols

$$\overline{G}_T(\Phi) > 0 \quad \Rightarrow \quad \overline{G}_{k-1,k}(\Phi)1_B > 0$$

for some time t_k and some state $B \in \mathcal{P}_{k-1}$. (The child subtrees are "atomic" in the sense that they involve only one time interval and only one parent state, but may involve more than one asset.) To prove this, let $k \geq 1$ be the smallest index for which $\overline{G}_k(\Phi) > 0$. If $k = 1$, then

$$\overline{G}_{0,1}(\Phi)1_\Omega > 0$$

and so there is arbitrage on the child subtree of the root node. Hence, we may assume that $k > 1$. Since the gain $\overline{G}_{k-1}(\Phi)$ is $\mathcal{P}_{k-1}$-measurable, for each $B \in \mathcal{P}_{k-1}$, the value of $\overline{G}_{k-1}(\Phi)$ is a constant c_B on B and so

$$0 \le \overline{G}_k(\Phi)1_B = \overline{G}_{k-1}(\Phi)1_B + \overline{G}_{k-1,k}(\Phi)1_B = c_B1_B + \overline{G}_{k-1,k}(\Phi)1_B$$

that is,

$$-c_B1_B \le \overline{G}_{k-1,k}(\Phi)1_B$$

Now let us examine the possibilities for the constants c_B as B ranges over $\mathcal{P}_{k-1}$.

If $c_B < 0$ for any $B \in \mathcal{P}_{k-1}$, then $\overline{G}_{k-1,k}(\Phi)1_B > 0$ and there is arbitrage on the child subtree of B. Thus, we may assume that $c_B \ge 0$ for all $B \in \mathcal{P}_{k-1}$. But then if $c_B > 0$ for any of the states $B \in \mathcal{P}_{k-1}$, we have $\overline{G}_{k-1}(\Phi) > 0$, which contradicts the definition of k. Hence, $c_B = 0$ for all $B \in \mathcal{P}_{k-1}$ and so $\overline{G}_{k-1}(\Phi) = 0$. It follows that

$$\overline{G}_{k-1,k}(\Phi) = \overline{G}_k(\Phi) > 0$$

and so at least one of the child subtrees at $B \in \mathcal{P}_{k-1}$ must satisfy

$$\overline{G}_{k-1,k}(\Phi)1_B > 0$$

that is, the child subtree at B exhibits arbitrage. Thus, in all cases, there is arbitrage on a child subtree of the model.

Theorem 5.3 *Let $\mathbb{M}$ be a discrete-time model. Then an arbitrage trading strategy Φ for $\mathbb{M}$ must exhibit arbitrage on some child subtree of the model; that is, the discounted gain from some parent state to each of its child states must be nonnegative and one such gain must be positive, in symbols*

$$\overline{G}_{k-1,k}(\Phi)1_B > 0$$

for some time t_k and some state $B \in \mathcal{P}_{k-1}$. $\square$

Martingale Measures

Arbitrage is an unfair condition in the market that is brought about by the unfair pricing of one or more assets. Hence, it should come as no surprise that a model will be arbitrage free if and only if its price processes are fair; that is, if and only if its price processes are martingales. Proof of this key fact, called the *First Fundamental Theorem of Asset Pricing*, is our next major goal.

Definition *Let $\mathbb{M}$ be a discrete-time model with state information structure*

$$\mathbb{F} = (\mathcal{P}_0, \dots, \mathcal{P}_T)$$

A probability measure $\mathbb{P}$ *on* Ω *is a* **martingale measure** (*also called a* **risk-neutral probability measure**) *for* $\mathbb{M}$ *if*
1) $\mathbb{P}$ *is strongly positive; that is,*

$$\mathbb{P}(\omega) > 0$$

for all $\omega \in \Omega$.
2) *For each asset* $\mathfrak{a}_j$, *the discounted price process* $(\overline{S}_{0,j}, \ldots, \overline{S}_{T,j})$ *is a martingale; that is, for all* $k \geq 0$,

$$\mathcal{E}_{\mathbb{P}}(\overline{S}_{k+1,j} \mid \mathcal{P}_k) = \overline{S}_{k,j} \qquad \square$$

We should emphasize that a martingale measure is a *theoretical* probability measure and generally has nothing whatever to do with the *natural probability* of a final state of the model, which is often determined from empirical data.

As a simple example, consider a one-period model and a "stock" whose initial price is 0 and whose final price is determined by flipping a coin. If the coin comes up heads, the price is 2 and if the coin comes up tails, the price is -1. Then assuming a risk-free rate of 0, the probability measure

$$\mathbb{P}(\text{heads}) = \frac{1}{3}, \quad \mathbb{P}(\text{tails}) = \frac{2}{3}$$

is a martingale measure, since the expected final price is 0. Note that this probability has absolutely nothing to do with the natural probability of the coin returning heads.

For convenience, let us recall some facts about martingales given in Theorem 4.2. First, if $\mathbb{P}$ is a martingale measure, then every discounted price process is fair for *every interval of time* $[t_k, t_{k+i}]$; that is,

$$\mathcal{E}(\overline{S}_{k+i,j} \mid \mathcal{P}_k) = \overline{S}_{k,j}$$

for any $i, k \geq 0$. Second, at any time t_i, the discounted *unconditional* expected price of any asset $\mathfrak{a}_j$ is the initial price of that asset; that is,

$$\mathcal{E}(\overline{S}_{k,j}) = S_{0,j}$$

Characterizing Martingale Measures

There are a number of ways to characterize martingale measures. Let $\mathbb{P}$ be a strongly positive probability measure.

First, we have the local martingale condition, which says that $\mathbb{P}$ is a martingale measure if and only if

$$\mathcal{E}(\overline{S}_{k+1,j} \mid B) = \overline{S}_{k,j}(B)$$

for all assets $\mathfrak{a}_j$, all times t_k and all states $B \in \mathcal{P}_k$, where $0 \le k < T$. In words, the expected time-t_{k+1} price $S_{k+1,j}$ of asset $\mathfrak{a}_j$ under the conditional probability measure $\mathbb{P}(\,\cdot\,\mid B)$ is just the time-t_k price of the asset on state B. As we will see later, this formula is useful in actually computing martingale measures.

In terms of gain, the local martingale condition is

$$\mathcal{E}(\Delta_{k,k+1}(\overline{\mathbb{S}}_j) \mid B) = 0$$

However, Example 5.3 shows that

$$\mathcal{E}(\overline{G}_T(\Phi[\mathfrak{a}_j, t_k, B])) = \mathcal{E}(\Delta_{k,k+1}(\overline{\mathbb{S}}_j)1_B) = \mathbb{P}(B)\mathcal{E}(\Delta_{k,k+1}(\overline{\mathbb{S}}_j) \mid B)$$

and so the local martingale condition is equivalent to the condition that the expected gain of all atomic trading strategies is 0, in symbols,

$$\mathcal{E}(\overline{G}_T(\Phi[\mathfrak{a}_j, t_k, B])) = 0 \tag{5.1}$$

for all assets $\mathfrak{a}_j$, all times t_k and all states $B \in \mathcal{P}_k$. This result characterizes the martingale condition, which is a condition involving *conditional* expectation, in terms of a statement involving only *unconditional* expectations.

Theorem 5.4 *Let* $\mathbb{M}$ *be a discrete-time model with state information structure*

$$\mathbb{F} = (\mathcal{P}_0, \dots, \mathcal{P}_T)$$

A strongly positive probability measure $\mathbb{P}$ *on* Ω *is a martingale measure for* $\mathbb{M}$ *if and only if the expected gain of every atomic trading strategy is* 0, *in symbols,*

$$\mathcal{E}(\overline{G}_T(\Phi[\mathfrak{a}_j, t_k, B])) = 0$$

for all assets $\mathfrak{a}_j$, *all times* t_k *and all states* $B \in \mathcal{P}_k$, *where* $0 \le k < T$. $\square$

According to the definition, a strongly positive probability measure $\mathbb{P}$ is a martingale measure if and only if all discounted price processes $\overline{\mathbb{S}}_j$ are martingales. In fact, $\mathbb{P}$ is a martingale measure if and only if all discounted value processes

$$(\overline{\mathcal{V}}_0(\Phi), \dots, \overline{\mathcal{V}}_T(\Phi))$$

for all self-financing trading strategies $\Phi = (\Theta_0, \dots, \Theta_T)$ are martingales.

One direction of the proof follows immediately from the fact that a discounted price process is a discounted value process. For the other direction, suppose that $\mathbb{P}$ is a martingale measure. Then for any value process $(\overline{\mathcal{V}}_0(\Phi), \dots, \overline{\mathcal{V}}_T(\Phi))$, since $\theta_{k+1,j}$ is $\mathcal{P}_k$-measurable, we have

$$\mathcal{E}(\overline{V}_{k+1}(\Phi) \mid \mathcal{P}_k) = \mathcal{E}\left(\sum_{j=1}^{n} \theta_{k+1,j} \overline{S}_{k+1,j} \,\Big|\, \mathcal{P}_k\right)$$

$$= \sum_{j=1}^{n} \mathcal{E}(\theta_{k+1,j} \overline{S}_{k+1,j} \mid \mathcal{P}_k)$$

$$= \sum_{j=1}^{n} \theta_{k+1,j} \mathcal{E}(\overline{S}_{k+1,j} \mid \mathcal{P}_k)$$

$$= \sum_{j=1}^{n} \theta_{k+1,j} \overline{S}_{k,j}$$

$$= \overline{V}_k(\Phi)$$

and so the discounted value process is a martingale.

Now, we can expand on this result as follows. If the discounted value process of any self-financing trading strategy Φ is a martingale, then Theorem 4.2 implies that

$$\mathcal{E}(\overline{V}_k(\Phi)) = \overline{V}_0(\Phi)$$

for all $0 < k \le T$. In terms of gains, since $\overline{V}_0(\Phi)$ is a constant, this is

$$\mathcal{E}(\overline{G}_k(\Phi)) = 0$$

for all $0 < k \le T$. Of course, this implies that the expected final gain is 0,

$$\mathcal{E}(\overline{G}_T(\Phi)) = 0$$

But we can now take Φ to be the atomic strategies $\Phi[\mathfrak{a}_j, t_k, B]$ and invoke Theorem 5.4 to show that $\mathbb{P}$ is a martingale measure. Hence, we have come full circle.

Theorem 5.5 *Let* $\mathbb{M}$ *be a discrete-time model with state information structure*

$$\mathbb{F} = (\mathcal{P}_0, \dots, \mathcal{P}_T)$$

The following are equivalent for a strongly positive probability measure $\mathbb{P}$.
1) $\mathbb{P}$ *is a martingale measure; that is,*

$$\mathcal{E}_{\mathbb{P}}(\Delta_{k,k+1}(\overline{\mathbb{S}}_j) \mid \mathcal{P}_k) = 0$$

   for all $j = 1, \dots, n$.
2) *The discounted value process* $(\overline{V}_0(\Phi), \dots, \overline{V}_T(\Phi))$ *of any self-financing trading strategy* Φ *is a martingale; that is,*

$$\mathcal{E}_{\mathbb{P}}(\overline{V}_{k+1}(\Phi) - \overline{V}_k(\Phi) \mid \mathcal{P}_k) = 0$$

   or, in terms of gain,

$$\mathcal{E}_{\mathbb{P}}(\overline{G}_{k,k+1}(\Phi) \mid \mathcal{P}_k) = 0$$

for all $k = 0, \ldots, T - 1$.

3) *The expected discounted cumulative gain of every self-financing trading strategy Φ and for every time t_k is 0; that is,*

$$\mathcal{E}_{\mathbb{P}}(\overline{G}_k(\Phi)) = 0$$

or equivalently, $\mathcal{E}_{\mathbb{P}}(\overline{\mathcal{V}}_k(\Phi)) = \overline{\mathcal{V}}_0(\Phi)$.

4) *The expected discounted final gain of every self-financing trading strategy Φ is 0; that is,*

$$\mathcal{E}_{\mathbb{P}}(\overline{G}_T(\Phi)) = 0$$

or equivalently, $\mathcal{E}_{\mathbb{P}}(\overline{\mathcal{V}}_T(\Phi)) = \overline{\mathcal{V}}_0(\Phi)$.

5) *The expected discounted final gain of every atomic trading strategy is 0; that is,*

$$\mathcal{E}_{\mathbb{P}}(\overline{G}_T(\Phi[\mathfrak{a}_j, t_k, B])) = 0$$

for all assets $\mathfrak{a}_j$, all times t_k and all states $B \in \mathcal{P}_k$, where $0 \leq k < T$.$\square$

Characterizing Arbitrage

We are now ready to prove the First Fundamental Theorem of Asset Pricing. For our proof, it helps to take a geometric approach by viewing a random variable not as a function, but as a vector. In particular, if we fix the order of the elements of Ω, say $\Omega = (\omega_1, \ldots, \omega_m)$, then any random variable $X: \Omega \to \mathbb{R}$ can be identified with its vector of values

$$X \leftrightarrow (X(\omega_1), \ldots, X(\omega_m))$$

We will use the same notion for the random variable X and the vector X. In this connection, we make the following definition.

Definition *Let $X = (x_1, \ldots, x_m)$ be a vector in $\mathbb{R}^m$.*
1) *X is **nonnegative**, written $X \geq 0$ if*

$$x_i \geq 0 \text{ for all } i = 1, \ldots, m$$

2) *X is **strictly positive**, written $X > 0$ if $X \geq 0$ and*

$$x_i > 0 \text{ for at least one } i = 1, \ldots, m$$

3) *X is **strongly positive**, written $X \gg 0$ if*

$$x_i > 0 \text{ for all } i = 1, \ldots, m$$ $\square$

Of course, a random variable X is nonnegative, strictly positive or strongly positive if and only if the corresponding vector X has this property. We also view a strongly positive probability measure $\mathbb{P}$ on Ω as a strongly positive **probability vector** $\Pi = (\pi_1, \ldots, \pi_m)$, where $\pi_i = \mathbb{P}(\omega_i)$ for $i = 1, \ldots, m$.

Theorem 5.6 (The First Fundamental Theorem of Asset Pricing) *A discrete-time pricing model* $\mathbb{M}$ *has no arbitrage opportunities if and only if it has a martingale measure.*

Proof. By definition, the model $\mathbb{M}$ has no arbitrage opportunities if and only if

$$\overline{G}_T(\Phi) \not> 0$$

for all self-financing trading strategies Φ. On the other hand, Theorem 5.5 says that a strongly positive probability measure $\mathbb{P}$ on Ω is a martingale measure if and only if

$$\mathcal{E}_{\mathbb{P}}(\overline{G}_T(\Phi)) = 0$$

for all self-financing trading strategies Φ. We must show that these two properties are equivalent.

One direction is pretty easy. If $\mathbb{M}$ has a martingale measure $\mathbb{P}$ and if

$$\overline{G}_T(\Phi) = (g_1, \ldots, g_m)$$

then

$$0 = \mathcal{E}_{\mathbb{P}}(\overline{G}_T(\Phi)) = g_1 \mathbb{P}(\omega_1) + \cdots + g_m \mathbb{P}(\omega_m)$$

and since $\mathbb{P}$ is strongly positive, it follows that $\overline{G}_T(\Phi)$ cannot be a strictly positive vector; that is, $\overline{G}_T(\Phi) \not> 0$. Hence, $\mathbb{M}$ has no arbitrage opportunities.

For the converse, we must work much harder. Assume that $\mathbb{M}$ has no arbitrage opportunities; that is, $\overline{G}_T(\Phi) \not> 0$ for all self-financing trading strategies Φ. Of course, since $\overline{G}_T(\Phi)$ is a random variable (or a vector), we *cannot* conclude that $\overline{G}_T(\Phi) \leq 0$, as we could with real numbers.

We wish to find a strongly positive probability vector $\Pi = (\pi_1, \ldots, \pi_m)$ for which $\mathcal{E}_{\Pi}(\overline{G}_T(\Phi)) = 0$ for all self-financing trading strategies Φ. Note that the expected value can be written as an inner product

$$\mathcal{E}_{\Pi}(\overline{G}_T(\Phi)) = \sum_{i=1}^{m} \overline{G}_T(\Phi)(\omega_i) \pi_i = \langle \overline{G}_T(\Phi), \Pi \rangle$$

so we seek a strongly positive probability vector Π for which

$$\langle \overline{G}_T(\Phi), \Pi \rangle = 0$$

Actually, it is sufficient to find a strongly positive *vector* $\Pi = (\pi_1, \ldots, \pi_m)$ for which this equation is valid, since we can divide each coordinate π_i by the sum $\sigma = \sum \pi_i$ to get a strongly positive *probability measure* with the same property.

Now, since the discounted gain function $\overline{G}_T$ is a linear transformation on the vector space of all self-financing trading strategies, it follows that the set $\mathcal{G}$ of all final gain vectors

$$\mathcal{G} = \{\overline{G}_T(\Phi) \mid \Phi \text{ is a trading strategy}\} \subseteq \mathbb{R}^m$$

is a subspace of $\mathbb{R}^m$. The absence of arbitrage condition $\overline{G}_T(\Phi) \not> 0$ is equivalent to the condition that this subspace intersects the **nonnegative orthant**

$$\mathbb{R}^m_+ = \{(x_1, \ldots, x_m) \mid x_i \geq 0\}$$

in $\mathbb{R}^m$ only at the origin; that is,

$$\mathcal{G} \cap \mathbb{R}^m_+ = \{0\}$$

Thus, in geometric terms, we must prove that if a subspace $\mathcal{G}$ of $\mathbb{R}^m$ intersects the nonnegative orthant trivially, then there is a nonzero vector Π in the first quadrant (and hence strongly positive) that is *orthogonal* to $\mathcal{G}$.

The case $m = 2$ is shown in Figure 5.5.

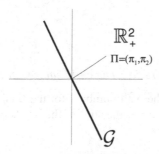

Figure 5.5: The case $\mathbb{R}^2$

In this figure, $\mathcal{G}$ is a one-dimensional subspace of $\mathbb{R}^2$ that intersects the first nonnegative quadrant trivially. It is pretty clear from the figure that there is a vector $\Pi = (\pi_1, \pi_2)$ in the first quadrant that is orthogonal to $\mathcal{G}$. To prove this, note that the subspace $\mathcal{G}$ is a line through the origin with slope $m < 0$. Hence, the line $y = -x/m$ is orthogonal to $\mathcal{G}$, since its slope is the negative reciprocal of the slope of $\mathcal{G}$. Hence, $\Pi = (m, -1)$ will do.

Unfortunately, this geometric argument does not generalize easily to larger dimensions. However, in Theorem A.7 of Appendix A, we prove that there does exist such a vector orthogonal to $\mathcal{G}$. The proof requires some facts from convexity theory, which are developed in that appendix.□

Computing Martingale Measures

Computing a martingale measure $\mathbb{P}$ for a discrete-time model (if it exists) is actually quite easy, since it amounts simply to solving a series of systems of linear equations. Suppose we have labeled the edges of the state tree for the model with positive real numbers. Figure 5.6 shows a child subtree with parent B_k and children

$$\{B_{k+1,1}, \ldots, B_{k+1,s_k}\}$$

Also, the edge label from B_k to $B_{k+1,i}$ is $p_{k+1,i}$.

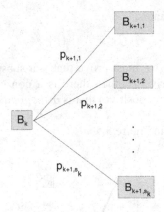

Figure 5.6: The child subtree at B_k

Theorem 4.1 implies that the path numbers for the final states form a probability distribution $\mathbb{P}$ on Ω if and only if the sum of the labels on each child subtree is 1; that is, if and only if

$$1 = \sum_{i=1}^{s_k} p_{k+1,i} \tag{5.2}$$

Moreover, in this case, the edge labels are the conditional probabilities

$$p_{k+1,i} = \mathbb{P}(B_{k+1,i} \mid B_k)$$

Therefore, the local martingale condition at B_k,

$$\mathcal{E}(\overline{S}_{k+1,j} \mid B_k) = \overline{S}_{k,j}(B_k)$$

can be written in the form

$$\overline{S}_{k,j}(B_k) = \sum_{i=1}^{s_k} \overline{S}_{k+1,j}(B_{k+1,i}) p_{k+1,i} \tag{5.3}$$

for each $j = 1, \ldots, m$. We will refer to equations (5.2) and (5.3) as the **local martingale equations** for a martingale measure.

Theorem 5.7 *Let* $\mathbb{M}$ *be a discrete time model. If the edges of the state tree are labeled with positive real numbers* $p_{k+1,i}$ *as described above, then the path numbers define a martingale measure* $\mathbb{P}$ *on* Ω *if and only if the local martingale equations (5.2) and (5.3) hold.* $\square$

We can use the local martingale equations to compute a martingale measure by working backward in time, as shown in the following example.

Example 5.4 The left half of Figure 5.7 shows a state tree, with stock prices for a two-asset model. We assume that the risk-free rates are 0.

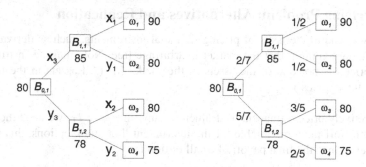

Figure 5.7: Computing martingale probabilities

For the state $B_{1,1}$, the local martingale equations are (using the simpler variable names x_i and y_i for the variables)

$$x_1 + y_1 = 1$$
$$90x_1 + 80y_1 = 85$$

whose solution is $x_1 = y_1 = 1/2$, as shown on the right in Figure 5.7. Similarly, for block $B_{1,2}$ we have

$$x_2 + y_2 = 1$$
$$80x_2 + 75y_2 = 78$$

with solution $x_2 = 3/5$, $y_2 = 2/5$. Finally, for the block $B_{0,1}$ we have

$$x_3 + y_3 = 1$$
$$85x_3 + 78y_3 = 80$$

with solution $x_3 = 2/7$, $y_3 = 5/7$. Thus, the martingale measure is given by the path numbers

$$\mathbb{P}(\omega_1) = \frac{2}{7} \cdot \frac{1}{2} = \frac{2}{14}$$
$$\mathbb{P}(\omega_2) = \frac{2}{7} \cdot \frac{1}{2} = \frac{2}{14}$$
$$\mathbb{P}(\omega_3) = \frac{5}{7} \cdot \frac{3}{5} = \frac{3}{7}$$
$$\mathbb{P}(\omega_4) = \frac{5}{7} \cdot \frac{2}{5} = \frac{2}{7}$$

Moreover, since this model has a martingale measure, the First Fundamental Theorem of Asset Pricing implies that the model is arbitrage-free.$\square$

The Pricing Problem: Alternatives and Replication

We now come to the issue of pricing financial instruments, such as derivatives, under the assumption that the market is arbitrage-free. Note that the instrument to be priced is not one of the assets in the model $\mathbb{M}$. (All assets in the model already have prices.)

To effectively price a financial instrument at any time $t_k < t_T$, we need the final time-t_T payoff random variable for the instrument. For stock options, this is not a problem, since the final payoff of a call option is

$$X = (S_T - K)^+$$

and the final payoff of a put option is

$$X = (K - S_T)^+$$

where K is the strike price. For example, in a two-state economy, suppose a_2 is a stock with initial price 100 and final price

$$S_T(\omega_1) = 120, \quad S_T(\omega_2) = 90$$

Then a call with strike price $K = 95$ has final payoff

$$X(\omega) = (S_T - K)^+(\omega) = \begin{cases} 25 & \text{if } \omega = \omega_1 \\ 0 & \text{if } \omega = \omega_2 \end{cases}$$

Actually, from the point of view of pricing a financial instrument, all that matters is the payoff random variable X—the precise nature of the instrument (call, put, uranium mine, etc.) is not relevant. Thus, we are really pricing *random variables* under the no-arbitrage assumption. When a random variable is thought of as a final payoff, it has a special name.

Definition *A random variable* $X \colon \Omega \to \mathbb{R}$ *is also called an* **alternative,** *or* **contingent claim.**$\square$

Note that some authors require that an alternative be nonnegative, the idea being that a claim based on an option will not have negative payoffs, since the claim will simply expire worthless. However, we do not make this additional restriction.

In some sense, any alternative $X: \Omega \to \mathbb{R}$ defines a "financial instrument" with final payoff X. In fact, we will assume that for any random variable $X: \Omega \to \mathbb{R}$, some investor will be willing to buy and some investor will be willing to sell a "financial instrument" whose final payoff is X. (Actually, this is one of the few assumptions about the market that we have made that is actually plausible.)

Thus, the pricing problem is the problem of pricing *alternatives*. The procedure for pricing an alternative X is first to find a self-financing trading strategy Φ within the model whose final payoff is equal to X; that is, for which

$$\mathcal{V}_T(\Phi) = X$$

Then we simply set the time-t_k price of X equal to the time-t_k price $\mathcal{V}_k(\Phi)$ of Φ. Indeed, *any other choice will lead to arbitrage* (when the alternative is added to the model). For if the time-t_k price P_k of X is not equal to $\mathcal{V}_k(\Phi)$, then an investor can enter the market at that time, buying the cheaper of Φ and X and selling the more expensive one. This produces a profit at time t_k and at the end the investor can liquidate his long position and use the proceeds to exactly pay off the short position.

This prompts the following definition.

Definition *Let* $\mathbb{M}$ *be a discrete-time model and let* $X: \Omega \to \mathbb{R}$ *be an alternative. A* **replicating trading strategy** *for* X *is a self-financing trading strategy* Φ *in* $\mathbb{M}$ *whose payoff is equal to* X; *that is, for which*

$$\mathcal{V}_T(\Phi) = X$$

An alternative X *that has at least one replicating strategy is said to be* **attainable***. The model* $\mathbb{M}$ *is* **complete** *if every alternative is attainable.*$\square$

The set $\mathcal{M}$ of all attainable alternatives is a subspace of the vector space $\mathrm{RV}(\Omega)$ of all random variables on Ω. We leave verification of this to the reader.

Computing a Replicating Trading Strategy

Finding a replicating trading strategy $\Phi = (\Theta_1, \ldots, \Theta_T)$, if it exists, for an alternative X is also just a matter of solving some systems of linear equations, working backward in time. The key observation is that for a self-financing trading strategy Φ, if we know the time-t_k liquidation value $\mathcal{V}_k(\Theta_k)$ then the equation

$$\langle S_k, \Theta_k \rangle = \mathcal{V}_k(\Theta_k) \tag{5.4}$$

can be used to compute the quantities Θ_k. Then we can compute the time-t_{k-1} acquisition value $\mathcal{V}_{k-1}(\Theta_k)$, which gives us the liquidation value $\mathcal{V}_{k-1}(\Theta_{k-1})$, since

$$\mathcal{V}_{k-1}(\Theta_{k-1}) = \mathcal{V}_{k-1}(\Theta_k)$$

Then the process can be repeated. Thus, starting with the known final liquidation value $\mathcal{V}_T(\Phi) = X$ and working backward in time, we can compute the entire trading strategy Φ. Let us look at an example.

Example 5.5 Figure 5.8 shows the state tree from Example 5.4. We assume that the risk-free rates are 0.

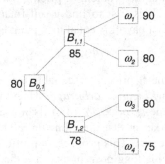

Figure 5.8: A state tree

Let us compute a self-financing trading strategy $\Phi = (\Theta_1, \Theta_2)$ that replicates the alternative X defined by

$$X(\omega_1) = 100, X(\omega_2) = 90, X(\omega_3) = 80, X(\omega_4) = 70$$

Thus, we want Θ_2 to satisfy

$$\mathcal{V}_2(\Theta_2)(\omega) = \begin{cases} 100 & \text{if } \omega = \omega_1 \\ 90 & \text{if } \omega = \omega_2 \\ 80 & \text{if } \omega = \omega_3 \\ 70 & \text{if } \omega = \omega_4 \end{cases}$$

The liquidation value equations (5.4) for the child subtree at $B_{1,1}$ are

$$\theta_{2,1}(\omega_1) + 90\theta_{2,2}(\omega_1) = 100$$
$$\theta_{2,1}(\omega_2) + 80\theta_{2,2}(\omega_2) = 90$$

However, Θ_2 is constant on $B_{1,1} = \{\omega_1, \omega_2\}$ so we may replace ω_2 by ω_1 in the second equation to get

$$\theta_{2,1}(\omega_1) + 90\theta_{2,2}(\omega_1) = 100$$
$$\theta_{2,1}(\omega_1) + 80\theta_{2,2}(\omega_1) = 90$$

which has the unique solution

$$\Theta_2(\omega_1) = \Theta_2(\omega_2) = (10, 1)$$

In a similar way, the equations for $B_{1,2}$ are

$$\theta_{2,1}(\omega_3) + 80\theta_{2,2}(\omega_3) = 80$$
$$\theta_{2,1}(\omega_4) + 75\theta_{2,2}(\omega_4) = 70$$

and the previewable condition gives

$$\theta_{2,1}(\omega_3) + 80\theta_{2,2}(\omega_3) = 80$$
$$\theta_{2,1}(\omega_3) + 75\theta_{2,2}(\omega_3) = 70$$

with solution

$$\Theta_2(\omega_3) = \Theta_2(\omega_4) = (-80, 2)$$

Thus,

$$\Theta_2(\omega_1) = \Theta_2(\omega_2) = (10, 1)$$
$$\Theta_2(\omega_3) = \Theta_2(\omega_4) = (-80, 2)$$

Now we can compute the time-t_1 values,

$$\mathcal{V}_1(\Theta_1)(\omega_1) = \mathcal{V}_1(\Theta_2)(\omega_1) = 10 + 85 \cdot 1 = 95$$
$$\mathcal{V}_1(\Theta_1)(\omega_3) = \mathcal{V}_1(\Theta_2)(\omega_3) = -80 + 78 \cdot 2 = 76$$

From the values of $\mathcal{V}_1(\Theta_1)$, we get the equations

$$\theta_{1,1}(\omega_1) + 85\theta_{1,2}(\omega_1) = 95$$
$$\theta_{1,1}(\omega_3) + 78\theta_{1,2}(\omega_3) = 76$$

But Θ_1 is constant on Ω, and so for any $\omega \in \Omega$

$$\theta_{1,1}(\omega) + 85\theta_{1,2}(\omega) = 95$$
$$\theta_{1,1}(\omega) + 78\theta_{1,2}(\omega) = 76$$

This system has solution

$$\Theta_1(\omega) = \left(-\frac{950}{7}, \frac{19}{7}\right)$$

which is a portfolio consisting of a short position (sale) of $950/7 \approx 135.71$ units of risk-free asset and a long position (purchase) of $19/7 \approx 2.71$ shares of stock, for an initial cost of

$$-\frac{950}{7} + 80 \cdot \frac{19}{7} = \frac{570}{7} \approx 81.43$$

Thus, for an initial cost of \$81.43, we can acquire a portfolio that will replicate X. Note that under some states we have a profit; under others a loss. This is expected in a model with no arbitrage. $\square$

The Law of One Price and the Pricing Functionals

It is clear that the replicating strategy procedure can only be used to price *attainable* alternatives. However, there is still one potential problem, and that is the problem of multiple replicating strategies for a given alternative having *different* initial values. However, the assumption of the absence of arbitrage implies that this cannot happen.

Definition *The* **Law of One Price** *is the statement that*

$$\mathcal{V}_T(\Phi_1) = \mathcal{V}_T(\Phi_2) \quad \Rightarrow \quad \mathcal{V}_k(\Phi_1) = \mathcal{V}_k(\Phi_2)$$

for all $0 \leq k \leq T$ *and for all trading strategies* Φ_1 *and* Φ_2. $\square$

The absence of arbitrage implies that the Law of One Price must hold and the Law of One Price ensures that the following pricing functionals are well-defined.

Definition *Let* $\mathbb{M}$ *be a model with no arbitrage. For any time* t_k, *the* **time-t_k pricing functional** $\mathcal{I}_k \colon \mathcal{M} \to \mathbb{R}$ *is defined on the vector space* $\mathcal{M}$ *of all attainable alternatives as follows: If* $X \in \mathcal{M}$, *then*

$$\mathcal{I}_k(X) = \mathcal{V}_k(\Phi)$$

for any replicating trading strategy Φ *for* X. *If* $k = 0$, *then* $\mathcal{I} = \mathcal{I}_0$ *is called the* **initial pricing functional.** $\square$

Therefore, to price an alternative X at time t_k, we first find a replicating trading strategy Φ and then set

$$\mathcal{I}_k(X) = \mathcal{V}_k(\Phi)$$

However, we can say more. According to Theorem 5.5, if $\mathbb{P}$ is a martingale measure for the model, then all discounted value processes are martingales; that is, for all times t_k,

$$\mathcal{E}(\overline{\mathcal{V}}_T(\Phi) \mid \mathcal{P}_k) = \overline{\mathcal{V}}_k(\Phi)$$

Hence,

$$\overline{\mathcal{I}}_k(X) = \overline{\mathcal{V}}_k(\Phi) = \mathcal{E}(\overline{\mathcal{V}}_T(\Phi) \mid \mathcal{P}_k) = \mathcal{E}(\overline{X} \mid \mathcal{P}_k)$$

where $\overline{X} = X / S_{T,1}$.

Theorem 5.8 *Let* $\mathbb{M}$ *be a discrete-time model with no arbitrage and let* $\mathbb{P}$ *be a martingale measure for* $\mathbb{M}$. *Let* X *be an attainable alternative.*
1)   *The discounted time-t_k price of* X *is*

$$\overline{\mathcal{I}}_k(X) = \mathcal{E}_{\mathbb{P}}(\overline{X} \mid \mathcal{P}_k)$$

where $\overline{X} = X/S_{T,1}$. In words, the discounted time-t_k price of X is simply the conditional expectation of $\overline{X}$ given $\mathcal{P}_k$ under the martingale measure $\mathbb{P}$.

2) *In particular, the initial price of X is just the expected discounted final value under the martingale measure,*

$$\mathcal{I}_0(X) = \mathcal{E}_{\mathbb{P}}(\overline{X}) \qquad \square$$

Example 5.6 Let us consider once again the alternative

$$X(\omega_1) = 100, X(\omega_2) = 90, X(\omega_3) = 80, X(\omega_4) = 70$$

from Example 5.5. A martingale measure for this model was given in Example 5.4 by

$$\mathbb{P}(\omega_1) = \frac{2}{14}, \quad \mathbb{P}(\omega_2) = \frac{2}{14}, \quad \mathbb{P}(\omega_3) = \frac{3}{7}, \quad \mathbb{P}(\omega_4) = \frac{2}{7}$$

and so Theorem 5.8 implies that

$$\begin{aligned}
\mathcal{I}_0(X) &= \mathcal{E}_{\mathbb{P}}(\overline{\mathcal{V}}_T(\Phi)) \\
&= 100 \cdot \frac{2}{14} + 90 \cdot \frac{2}{14} + 80 \cdot \frac{3}{7} + 70 \cdot \frac{2}{7} \\
&= \frac{570}{7} \\
&\approx 81.43
\end{aligned}$$

just as we found in Example 5.5, but with a lot less computation!$\square$

The previous example shows that the martingale measure can save a great deal of effort in pricing alternatives.

Uniqueness of Martingale Measures

We have seen that a discrete-time model $\mathbb{M}$ is arbitrage-free if and only if it has a martingale measure. It is natural to wonder if a model can have more than one martingale measure.

Recall that a model $\mathbb{M}$ is *complete* if every alternative in $\mathbb{R}^m$ is attainable; that is, if for every $X \in \mathbb{R}^m$ there is a self-financing trading strategy Φ such that

$$\mathcal{V}_T(\Phi) = X$$

We will use the following fact from linear algebra. Any *strongly positive* probability distribution $\Pi = (\pi_1, \ldots, \pi_m)$ on Ω, where $\mathbb{P}_{\Pi}(\omega_k) = \pi_k$ defines an inner product on the vector space $\mathbb{R}^m$ by

$$\langle X, Y \rangle_{\Pi} = \sum_{i=1}^{m} x_i y_i \pi_i$$

Observe also that if $\mathbf{1} = (1, \ldots, 1)$, then for any vector (random variable) X,

$$\langle X, \mathbf{1} \rangle_\Pi = \sum_{i=1}^{m} x_i \pi_i = \mathcal{E}_\Pi(X)$$

and so the inner product gives the expected value. Now we can resolve the issue at hand.

Theorem 5.9 (The Second Fundamental Theorem of Asset Pricing) *Let* $\mathbb{M}$ *be a model with no arbitrage opportunities, and hence at least one martingale measure. Then there is a* unique *martingale measure on* $\mathbb{M}$ *if and only if the model* $\mathbb{M}$ *is complete.*
Proof. Suppose first that $\mathbb{M}$ is complete and that Π_1 and Π_2 are martingale measures on $\mathbb{M}$. We want to show that $\Pi_1 = \Pi_2$. Theorem 5.5 implies that

$$\mathcal{E}_{\Pi_1}(\overline{\mathcal{V}}_T(\Phi)) = \overline{\mathcal{V}}_0(\Phi) = \mathcal{E}_{\Pi_2}(\overline{\mathcal{V}}_T(\Phi))$$

and since the discounting periods are the same, we have

$$\mathcal{E}_{\Pi_1}(\mathcal{V}_T(\Phi)) = \mathcal{E}_{\Pi_2}(\mathcal{V}_T(\Phi))$$

But since $\mathbb{M}$ is complete, all random variables on Ω have the form $\mathcal{V}_T(\Phi)$ for some self-financing trading strategy Φ and so

$$\mathcal{E}_{\Pi_1}(X) = \mathcal{E}_{\Pi_2}(X)$$

for all random variables X on Ω. Taking any $\omega \in \Omega$ and setting $X = 1_{\{\omega\}}$ gives

$$\mathbb{P}_{\Pi_1}(\omega) = \mathbb{P}_{\Pi_2}(\omega)$$

which implies that $\Pi_1 = \Pi_2$. Thus, the martingale measure on $\mathbb{M}$ is unique.

For the converse, suppose that Π is a martingale measure on $\mathbb{M}$ and that the market is not complete. We want to find a different martingale measure Π^* on $\mathbb{M}$. Since $\mathbb{M}$ is not complete, the vector space $\mathcal{M}$ of all attainable vectors is a *proper* subspace of $\mathbb{R}^m$.

Let us consider the inner product defined on $\mathbb{R}^m$ by the martingale measure Π. Since $\mathcal{M}$ is a proper subspace of $\mathbb{R}^m$, there is a vector $Z = (z_1, \ldots, z_m)$ that is orthogonal to $\mathcal{M}$; that is, for any attainable vector $X = (x_1, \ldots, x_m)$, we have

$$\langle X, Z \rangle_\Pi = \sum_{i=1}^{m} x_i z_i \pi_i = 0$$

Since the vector $\mathbf{1} = (1, \ldots, 1)$ is attainable (just buy $1/S_{T,1}$ units of the risk-free asset and roll it over), we have

$$0 = \langle \mathbf{1}, Z \rangle_{\Pi} = \sum_{k=1}^{m} z_k \pi_k$$

Now consider the vector $\Pi^* = (\pi_1^*, \ldots, \pi_m^*)$ where

$$\pi_k^* = \pi_k(1 + cz_k)$$

and $c \in \mathbb{R}$. We can arrange it so that Π^* is strongly positive ($\pi_k^* > 0$ for all k) by choosing c small enough. Also,

$$\sum_{k=1}^{m} \pi_k^* = \sum_{k=1}^{m} \pi_k + c \sum_{k=1}^{m} z_k \pi_k = 1$$

and so Π^* is a strongly positive probability measure. Moreover, for any attainable vector $X \in \mathcal{M}$,

$$\begin{aligned}
\mathcal{E}_{\Pi^*}(X) &= \sum_{i=1}^{m} x_i \pi_i^* \\
&= \sum_{i=1}^{m} x_i(\pi_i + cz_i\pi_i) \\
&= \sum_{i=1}^{m} x_i\pi_i + c\sum_{i=1}^{m} x_i z_i \pi_i \\
&= \mathcal{E}_{\Pi}(X)
\end{aligned}$$

Hence, since the attainable vectors are precisely the vectors $\overline{V}_T(\Phi)$, we can say that for any self-financing trading strategy Φ,

$$\mathcal{E}_{\Pi^*}(\overline{V}_T(\Phi)) = \mathcal{E}_{\Pi}(\overline{V}_T(\Phi))$$

and since Π is a martingale measure, Theorem 5.5 implies that

$$\mathcal{E}_{\Pi^*}(\overline{V}_T(\Phi)) = \mathcal{E}_{\Pi}(\overline{V}_T(\Phi)) = \overline{V}_0(\Phi)$$

But this same theorem then tells us that Π^* is also a martingale measure.$\square$

Exercises

1. Prove that a model may have arbitrage even though the atomic trading strategies $\Phi[a_j, t_k, B]$ do not exhibit such behavior.
2. For the state tree in Figure 5.8, compute a self-financing trading strategy $\Phi = (\Theta_1, \Theta_2)$ that replicates the alternative

$$X(\omega_1) = 95, X(\omega_2) = 90, X(\omega_3) = 85, X(\omega_4) = 75$$

   Assume that the risk-free rates are 0.
3. For the state tree in Figure 5.9,

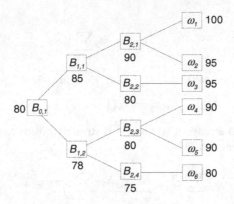

Figure 5.9

replicate the alternative

$$(100, 100, 95, 90, 90, 85)$$

Assume that the risk-free rates are 0. *Hint*: There is more than one possible answer.

4. Suppose that some derivative asset (such as an option) is not priced at its fair price at some time t_k in some state $B \in \mathcal{P}_k$. Explain how to take advantage of this opportunity to make a profit.

5. Consider the following game. Three fair coins are flipped. The player wins if three heads occur, otherwise the casino wins. For every \$0.25 the player wagers, the casino must put up \$2.00, making the wager fair. Imagine now that the casino wants to hedge its position against a player who wishes to wager \$1 million. (The casino is at risk for \$8 million.) Accordingly, the casino finds a "market maker" in coin-tossing bets and does the following: Before the first toss, it bets \$1 million on heads at even money; before the second toss (if there is one), it bets \$2 million on heads at even money and before the third toss (if there is one), it bets \$4 million on heads at even money. Track the value of the casino's and the player's portfolio during the game. Justify the statement that the casino has entered into a self-financing, replicating complete hedge.

6. Prove that the set $\mathcal{T}$ of all *self-financing* trading strategies is a vector space under the operations of coordinate-wise addition

$$(\Theta_{1,1}, \dots, \Theta_{1,T}) + (\Theta_{2,1}, \dots, \Theta_{2,T}) = (\Theta_{1,1} + \Theta_{2,1}, \dots, \Theta_{1,T} + \Theta_{2,T})$$

and scalar multiplication

$$a(\Theta_1, \dots, \Theta_T) = (a\Theta_1, \dots, a\Theta_T)$$

7. Consider the self-financing trading strategy

$$\Phi = (\Theta_1, \dots, \Theta_T)$$

where

$$\Theta_i = (\theta_{i,1}, \dots, \theta_{i,n})$$

For any nonzero real number a, let

$$\Phi' = (\Theta'_1, \dots, \Theta'_T)$$

where

$$\Theta'_i = (\theta_{i,1} + a1_\Omega, \dots, \theta_{i,n})$$

Show that Φ' is self-financing.

8. Prove that the set $\mathcal{M}$ of all attainable alternatives is a subspace of the vector space $RV(\Omega)$ of all random variables on Ω.

9. Prove that the Law of One Price is equivalent to the following: For all trading strategies Φ,

$$\mathcal{V}_T(\Phi) = 0 \quad \Rightarrow \quad \mathcal{V}_0(\Phi) = 0$$

10. Consider a model $\mathbb{M}$ with two assets: the risk-free asset and a stock. If the risk-free rates r_k are large enough, will there always be an arbitrage opportunity? Explain your answer. Does this apply to models with more than one risky asset?

11. Consider the following game. A set of 3 coins exists. The first coin is fair, the second coin has probability of heads equal to 0.55 and the third coin has probability of heads 0.45. Draw a state tree indicating the possible outcomes along with their probabilities. Find the path-weight probability distribution.

12. Show that the replicating relation defined by $\Phi_1 \equiv \Phi_2$ if and only if Φ_1 replicates Φ_2 is an *equivalence relation* on the set of self-financing trading strategies; that is, the relation satisfies the following conditions:
    a) (**reflexivity**) $\Phi_1 \equiv \Phi_1$
    b) (**symmetry**) $\Phi_1 \equiv \Phi_2$ implies $\Phi_2 \equiv \Phi_1$
    c) (**transitivity**) $\Phi_1 \equiv \Phi_2$ and $\Phi_2 \equiv \Phi_3$ implies $\Phi_1 \equiv \Phi_3$

13. Prove that if any *strictly positive* alternative is attainable then the market is complete.

14. A self-financing trading strategy Φ is **admissible** if its value at all times is nonnegative; that is, $\mathcal{V}_i(\Phi) \geq 0$ for all $i = 0, \dots, T$. Prove that if a model $\mathbb{M}$ has an arbitrage trading strategy, then it also has an admissible arbitrage trading strategy.

A Single-Period, Two-Asset, Two-State Model

Consider a simple single-period, two-asset, two-state model $\mathbb{M}$. The model has two assets $\mathcal{A} = (\mathfrak{a}_1, \mathfrak{a}_2)$ where $\mathfrak{a}_1$ is the risk-free bond at rate r and $\mathfrak{a}_2$ is an underlying stock with initial price S_0 and final price S_T. The model has only two states of the economy; that is, $\Omega = \{\omega_1, \omega_2\}$. It is customary to express the final stock price in terms of the initial price. In state ω_1 the stock price is multiplied by a factor u so that

$$S_T = S_0 u$$

and in state ω_2 the price is multiplied by a factor d so that

$$S_T = S_0 d$$

We will assume that $d < u$. The following exercises pertain to this model.

15. Show that $\mathbb{M}$ is complete if and only if $d < u$.
16. Consider an option with payoff X given by

$$X(\omega_1) = f_u$$
$$X(\omega_2) = f_d$$

Find a replicating portfolio for X.
17. Find the initial price of X in the previous exercise.
18. Set

$$\pi = \frac{e^{rT} - d}{u - d}$$

and show that the price of the derivative is

$$e^{-rT}[\pi f_u + (1 - \pi)f_d]$$

What does this tell you about $(\pi, 1 - \pi)$?
19. Show that there is no arbitrage in this model if and only if $d < e^{rT} < u$.
20. A day trader is interested in a particular stock currently priced at $100. His assessment is that by the end of the day, the stock will be selling for either $101 or $99. A European call is available at a strike price of $99.50. How should it be priced? Assume that $r = 4\%$.
21. a)   Suppose a certain security is currently selling for 160. At time T the security will sell for either $200 or $140. Price a European put on this asset with strike price $180, assuming no arbitrage and interest rate $r = 0$.
    b)   Suppose you are fortunate enough to acquire the put described above for only $20. Describe the various portfolios that include the put that will guarantee a profit.

A Single-Period, Two-Asset, Three-State Model

Consider now a single-period, two-asset model with three states. Assume a risk-free rate of 0. Suppose that

$$S_{0,2} = 25$$

and

$$S_{1,2}(\omega_1) = 40, S_{1,2}(\omega_2) = 30, S_{1,2}(\omega_3) = 20$$

22. Show that the model is not complete.

23. Find all martingale measures for this model.
24. Show that the following are martingale measures:

$$\Pi_1 = \left(\frac{1}{12}, \frac{4}{12}, \frac{7}{12} \right)$$

$$\Pi_2 = \left(\frac{1}{6}, \frac{1}{6}, \frac{4}{6} \right)$$

25. Find a replicating trading strategy (portfolio) and price a call option with strike price 20 using the two martingale measures of the previous exercise.

Chapter 6
The Binomial Model

In this chapter, we discuss a specific discrete-time model known as the **binomial model**, first described in 1979 by John Cox, Stephen Ross and Mark Rubinstein. In Chapter 10, we will use this model to derive the famous Black–Scholes option pricing formula.

The General Binomial Model

Here are the details of the general binomial model, which we denote by $\mathbb{M}$.

Times

The **lifetime** of a binomial model is a positive number L, which is divided into T time intervals

$$t_0 < t_1 < \cdots < t_T$$

of equal length Δt; that is, $t_i - t_{i-1} = \Delta t$. Thus,

$$L = t_T - t_0 = T\Delta t$$

Assets

The binomial model has only two assets: The risk-free asset $\mathfrak{a}_1$ and a risky asset $\mathfrak{a}_2$, which we will generally think of as a stock for concreteness.

The States of the Economy

The binomial model assumes that during each time interval $[t_k, t_{k+1}]$, the economy either goes up, called an **up-tick** in the economy or it goes down, called a **down-tick** in the economy. Moreover, each movement is independent of previous movements. We will denote an up-tick in the economy by U and a down-tick by D. Thus, the **state space** for the model is the set

$$\Omega = \Omega_T = \{U, D\}^T$$

of all strings of U's and D's of length T. These are the **final states** of the economy. We also let $\Omega_k = \{U, D\}^k$ be the set of all strings of U's and D's of length k. Figure 6.1 shows a portion of the state tree for the binomial model.

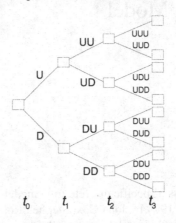

Figure 6.1: State tree for a binomial model

We denote the prefix of any $\omega \in \Omega$ of length k by $[\omega]_k$. For each $\delta = e_1 \cdots e_k \in \Omega_k$, the **intermediate state** $B_\delta \in \mathcal{P}_k$ is the set of all final states having prefix δ; that is,

$$B_\delta = \{\omega \in \Omega \mid [\omega]_k = \delta\}$$

Thus, $\mathcal{P}_k$ has exactly 2^k blocks. For example, if $T = 4$, then $\mathcal{P}_2$ consists of the four intermediate states B_{UU}, B_{UD}, B_{DU} and B_{DD}. For instance,

$$B_{UU} = \{\underline{UU}UU, \underline{UU}UD, \underline{UU}DU, \underline{UU}DD\}$$

(We have underlined the prefix in each case.) At time t_0, there is only one (initial) state $B_\epsilon = \Omega$. This corresponds to the empty string ϵ, which is a prefix of all strings.

Each block $B_\delta \in \mathcal{P}_k$ has exactly two children $B_{\delta U}$ and $B_{\delta D}$ and so a tree with this property is called a **binary tree**. This also accounts for the name *binomial model*.

The Natural Probabilities

For each time interval $[t_k, t_{k+1}]$, there is a **natural probability** p_k of an up-tick in the economy as well as a natural probability $1 - p_k$ of a down-tick.

The Price Functions

The time-t_k price of the stock is denoted by S_k, which is a random variable on Ω. If the economy has an up-tick during the interval $[t_k, t_{k+1}]$, the stock price rises by a factor of $u_k \geq 1$ to $S_{k+1} = S_k u_k$. If the economy has a down-tick, the stock price falls by a factor of $0 < d_k \leq 1$ to $S_{k+1} = S_k d_k$. We call u_k the time-t_k **up-tick factor**, d_k the time-t_k **down-tick factor**. The values u_k and d_k are called the **tick parameters** of the model. Note that

$$0 < d_k \leq 1 \leq u_k$$

We also require that an up-tick followed by a down-tick be the same as a down-tick followed by an up-tick, in symbols,

$$u_k d_{k+1} = d_k u_{k+1}$$

or equivalently,

$$\frac{u_{k+1}}{u_k} = \frac{d_{k+1}}{d_k}$$

for all k. Trees with this property are said to be **recombining**.

If $\omega \in \Omega$, we let $\delta_k(\omega)$ be the product of the up-tick and down-tick factors that determine the time-t_k price S_k. For example, if the final state of the economy is $\omega = UUDUDDU$, then $\delta_5(\omega) = u_0 u_1 d_2 u_3 d_4$ and so

$$S_5(\omega) = S_0 u_0 u_1 d_2 u_3 d_4 = S_0 \delta_5(\omega)$$

In general, we have

$$S_k(\omega) = S_0 \delta_k(\omega)$$

for any $\omega \in \Omega$ and in particular, the final price is

$$S_T(\omega) = S_0 \delta_T(\omega)$$

The fact that S_k is $\mathcal{P}_k$-measurable is reflected in the fact that the value $S_k(\omega)$ depends only on the prefix $[\omega]_k$ of ω and thus only on what has happened up to time t_k.

The price of the risk-free asset is, as always, given by the risk-free rates. As usual, we let r_k be the risk-free rate for the interval $[t_k, t_{k+1}]$.

Martingale Measures in the Binomial Model

Let us now compute a martingale measure $\mathbb{P}$ for a binomial model. We use the local martingale condition as described in the previous chapter. Figure 6.2 shows a child subtree for the time interval $[t_k, t_{k+1}]$ in the state tree of a binomial model.

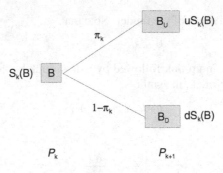

Figure 6.2

The local martingale equations that define a martingale measure are

$$u\overline{S}_k(B)\pi_k + d\overline{S}_k(B)(1 - \pi_k) = \overline{S}_k(B)$$

where the overbars on the left mean divide by $S_{k+1,1}$ and the overbar on the right means divide by $S_{k,1}$. Hence,

$$uS_k(B)\pi_k + dS_k(B)(1 - \pi_k) = e^{r_k\Delta t}S_k(B)$$

which simplifies to

$$u\pi_k + d(1 - \pi_k) = e^{r_k\Delta t}$$

(We can still use this equation to define π even if $S_k(B) = 0$.) Hence, the *unique* martingale measure probability is given by

$$\pi_k = \frac{e^{r_k\Delta t} - d_k}{u_k - d_k}$$

provided that it is also strongly positive; that is, provided that $0 < \pi_k < 1$ for all k or equivalently,

$$d_k < e^{r_k\Delta t} < u_k$$

for all k.

Theorem 6.1 *The binomial model is free of arbitrage if and only if*

$$d_k < e^{r_k\Delta t} < u_k$$

for all $k = 0, \ldots, T - 1$. In this case, the model is complete and the unique martingale measure $\mathbb{P}$ on $\mathbb{M}$ is defined by the path numbers in the state tree when the up-tick edges of the tree are labeled with the **martingale up-tick probabilities**

$$\pi_k = \frac{e^{r_k\Delta t} - d_k}{u_k - d_k}$$

and the down-tick edges are labeled with the **martingale down-tick probabilities** $1 - \pi_k$ *as shown in Figure 6.2..*□

Pricing in the Binomial Model

Since the binomial model is arbitrage-free and complete, the pricing functionals $\mathcal{I}_k$ are well defined. In particular, for any random variable X on Ω, Theorem 5.8 implies that

$$\mathcal{I}_0(X) = e^{-r\Delta t}\mathcal{E}(X)$$

where $r = \sum r_k$ is the sum of the risk-free rates and the expected value is taken under the martingale measure.

Standard Binomial Models

A binomial model is **standard** if the following hold:

1)   the up-tick probabilities $u = u_k$ are the same for all times
2)   the down-tick probabilities $d = d_k$ are the same for all times
3)   the risk-free rate $r = r_k$ is the same for all times
4)   the natural probabilities are the same for all times

The models that we have described above are sometimes called **flexible models** to distinguish them from standard models. The terms standard and flexible also apply to the state tree of a binomial model.

We summarize the properties of a standard model in the following theorem. Let

$$N_U(\omega) = \text{number of } U\text{'s in } \omega$$
$$N_D(\omega) = \text{number of } D\text{'s in } \omega$$

Theorem 6.2 *The standard binomial model is free of arbitrage if and only if*

$$d < e^{r\Delta t} < u$$

In this case, the time-t_k stock price function S_k is given by

$$S_k(\omega) = S_0 u^{N_U([\omega]_k)} d^{N_D([\omega]_k)}$$

for any $\omega \in \Omega$. In particular, the final price is

$$S_T(\omega) = S_0 u^{N_U(\omega)} d^{N_D(\omega)}$$

Moreover, the model is complete and the unique martingale measure $\mathbb{P}$ on $\mathbb{M}$ is defined by

$$\mathbb{P}(\omega) = \pi^{N_U(\omega)}(1 - \pi)^{N_D(\omega)}$$

for any $\omega \in \Omega_T$, where

$$\pi = \frac{e^{r\Delta t} - d}{u - d}$$

is the **martingale up-tick probability.** $\square$

Pricing in a Standard Binomial Model

The pricing functional has a nice form in a standard model,

$$\begin{aligned}
\mathcal{I}_0(X) &= e^{-rL}\mathcal{E}(X) \\
&= e^{-rL}\sum_{\omega\in\Omega} X(\omega)\mathbb{P}(\omega) \\
&= e^{-rL}\sum_{\omega\in\Omega} X(\omega)\pi^{N_U(\omega)}(1-\pi)^{N_D(\omega)}
\end{aligned}$$

For a call option, the final payoff is

$$X(\omega) = (S_T(\omega) - K)^+ = (S_0 u^{N_U(\omega)} d^{N_D(\omega)} - K)^+$$

But this depends on ω only through $N_U(\omega)$, since $N_D(\omega) = T - N_u(\omega)$ and so we can regroup the terms in the final summation based on the value of $N_U(\omega)$. Hence, for each of the $\binom{T}{k}$ sequences $\omega \in \Omega$ with exactly k U's, we have

$$X(\omega) = (S_0 u^k d^{T-k} - K)^+$$

Thus, we have a nice formula for the price of a European call option under a complete standard binomial model. A similar argument works for put options.

Theorem 6.3 *Let* $\mathbb{M}$ *be a complete standard binomial model with no arbitrage. Then a European call option with strike price* K *expiring at the end of the model has initial value*

$$\mathcal{I}_0(Call) = e^{-rL}\sum_{k=0}^{T} \binom{T}{k}(S_0 u^k d^{T-k} - K)^+\pi^k(1-\pi)^{T-k}$$

and a European put option has initial value

$$\mathcal{I}_0(Put) = e^{-rL}\sum_{k=0}^{T} \binom{T}{k}(K - S_0 u^k d^{T-k})^+\pi^k(1-\pi)^{T-k}$$

where

$$\pi = \frac{e^{r\Delta t} - d}{u - d}$$

is the martingale up-tick probability. $\square$

Example 6.1 A certain stock is currently selling for 100. The feeling is that for each month over the next 2 months, the stock's price will rise by 1% or fall by

1%. Assuming a risk-free rate of 1%, calculate the price of a European call with the various strike prices $K = 102, K = 101, \; K = 100, K = 99, K = 98$ and $K = 97$.

Solution The martingale probability is

$$\pi = \frac{e^{r\Delta t} - d}{u - d} = \frac{e^{(0.01)(1/12)} - 0.99}{0.02} \approx 0.54$$

and so

$$\mathcal{I}_0 = e^{-0.01/6} \sum_{k=0}^{2} \binom{2}{k} (100(1.01)^k (0.99)^{2-k} - K)^+ (0.54)^k (0.46)^{2-k}$$
$$= 0.9983[0.2116(98.01 - K)^+ + 0.4968(99.99 - K)^+$$
$$+ 0.2916(102.01 - K)^+]$$

Thus, some calculation gives the following table:

K	S_0
102	0.0029
101	0.2959
100	0.5888
99	1.3725
98	2.1632
97	3.1615

□

Choosing the Tick Parameters in a Standard Binomial Model

Theorem 6.3 gives formulas for the prices of options under a binomial model. Of course, these formulas involve the tick parameters u and d, as well as the risk-free rate r. Notwithstanding the problem that these parameters are assumed to be constant throughout the model—an assumption that is not very reasonable—we still face the issue of how to choose the best values for these parameters to use in our pricing formulas.

We can get some handle on the risk-free rate r by using U.S. Treasury securities of an appropriate maturity or some other asset that is generally considered risk free, such as federally insured bank accounts. Of course, different risk-free assets will have different rates of return, which is a problem.

As to the issue of estimating the tick parameters u and d, one of the few *actual* values to which we have access in building a model are historical asset prices. Using these prices, we can estimate future trends through statistical means. For example, we can assume that the expected value and variance for a stock's price over the future life of the model are approximately equal to these same values for the relatively recent past. In order to relate these quantities to the choices for u

and d, we must take a closer look at the movement in a stock's price under a binomial model.

Let p be the natural (not the martingale) probability of an up-tick in the market. During each time interval, the stock price takes either an up-tick or a down-tick. Hence, we can define independent Bernoulli random variables $E_1, \ldots, E_T$ to track these growth factors by

$$\mathbb{P}(E_k = u) = p$$
$$\mathbb{P}(E_k = d) = 1 - p$$

Then the stock price at time t_k is given by

$$S_k = S_0 E_1 \cdots E_k$$

and the final time-t_T price is

$$S_T = S_0 E_1 \cdots E_T$$

Since

$$\frac{S_{k+1}}{S_k} = E_{k+1}$$

we will refer to E_{k+1} as the **simple return** of the stock price over the time period $[t_k, t_{k+1}]$. We can convert the simple return E_{k+1} into an *annualized instantaneous rate* of return s_{k+1} by solving the equation

$$E_{k+1} = e^{s_{k+1} \Delta t}$$

Thus, we define the (annualized) **instantaneous return** of the stock price for the interval $[t_k, t_{k+1}]$ to be

$$s_{k+1} = \frac{1}{\Delta t} \log E_{k+1}$$

To make the stock price look like exponential growth, we write

$$S_T = S_0 E_1 \cdots E_T = S_0 e^{\sum \log E_i} = S_0 e^{H_T}$$

where

$$H_T = \log\left(\frac{S_T}{S_0}\right) = \sum_{i=1}^{T} \log E_i$$

is called the **logarithmic growth** of the stock price.

Now, the expected value μ and variance s^2 of the instantaneous return are given by

$$\mu = \frac{1}{\Delta t}\mathcal{E}(\log E_i) = \frac{1}{\Delta t}(p\log u + q\log d)$$

$$s^2 = \frac{1}{(\Delta t)^2}\text{Var}(\log E_i) = \frac{1}{(\Delta t)^2}pq(\log u - \log d)^2$$

The constant μ is called the **drift** and the constant

$$\sigma = s\sqrt{\Delta t} = \frac{1}{\sqrt{\Delta t}}\sqrt{pq}(\log u - \log d)$$

is called the **local volatility** of the stock price. (The quantity s is called the **instantaneous volatility** of the stock price. The explanation for considering σ rather than s lies in the definition of Brownian motion given later in the book. In any case, the reader need not worry about this issue now.) Thus, we have

$$\mathcal{E}(\log E_i) = \mu\Delta t \quad \text{and} \quad \text{Var}(\log E_i) = \sigma^2 \Delta t$$

and so we can standardize $\log E_i$ to get (since $\sigma \neq 0$)

$$X_i = \frac{\log E_i - \mu\Delta t}{\sigma\sqrt{\Delta t}}$$

which are independent standard Bernoulli random variables with

$$X_i = \begin{cases} \frac{q}{\sqrt{pq}} & \text{with probability } p \\ \frac{-p}{\sqrt{pq}} & \text{with probability } q \end{cases}$$

To write the logarithmic growth in terms of the random variables X_i, we have

$$H_T = \sum_{i=1}^{T}\log E_i$$

$$= \sum_{i=1}^{T}(\mu\Delta t + \sigma\sqrt{\Delta t}X_i)$$

$$= \mu L + \sigma\sqrt{\Delta t}\sum_{i=1}^{T}X_i$$

This formula expresses the logarithmic growth as a sum of a *deterministic part* μL, which is a constant multiple of the lifetime L of the model and a *random part*

$$Q = \sigma\sqrt{\Delta t}\sum_{i=1}^{T}X_i$$

The stock price itself is given by

$$S_T = S_0 e^{H_T} = S_0 e^{\mu L + Q} = S_0 e^{\mu L + \sigma \sqrt{\Delta t} \sum_{i=1}^{T} X_i}$$

Thus, the drift defines the **deterministic factor** $e^{\mu L}$ in the stock price and shows that the stock price behaves in part like a risk-free asset with risk-free rate μ. Put another way, the deterministic term μL accounts for a steady positive change (drift) in the stock's price (if $\mu \neq 0$). The volatility determines the **random factor** e^Q.

One final note. We have seen that the **expected instantaneous return** is

$$\mu = \frac{1}{\Delta t} \mathcal{E}(\log E_i) = \frac{1}{\Delta t}(p \log u + q \log d)$$

On the other hand, the expected return $pu + qd$ can be expressed as an annualized instantaneous rate ν defined by

$$pu + qd = \epsilon^{\nu \Delta t}$$

that is,

$$\nu = \frac{1}{\Delta t} \log(pu + qd)$$

We call ν the **instantaneous expected return** of the stock price (note the subtle difference in terminology).

Random Walks

The sequence $(X_1, \ldots, X_T)$ that describes the random behavior of the logrithmic growth of the stock price over each subinterval is an example of a *random walk*. To understand random walks, imagine a flea that is constrained to jump along a straight line, say the x-axis. The flea starts at the point $x = 0$ at time $t = 0$ and during each interval of time (of length Δt) jumps randomly a distance a to the right or to the left. Assume that the probability of a jump to the right is p. This is shown in Figure 6.3.

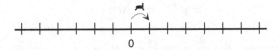

$$0$$

Figure 6.3: The random walk of a flea

Each variable X_i in the sequence (X_i) describes a single step in the flea's perambulations and the partial sums

$$U_k = \sum_{i=1}^{k} X_i$$

represent the position of the flea at time t_k.

Figure 6.4 shows two computer-generated random walks with $p = q = 1/2$. (These are called *symmetric random walks*.) As is customary in order to see the path clearly, each position of the flea is marked by a point in the plane, where the x-axis represents time and the y-axis represents position.

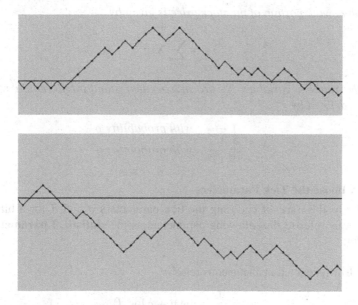

Figure 6.4: Random walks

There are many formulations of the random walk scenario, involving for example, drunks who are walking randomly along a street, gamblers playing a game of chance or the price of a stock.

Let us summarize what we have learned about the binomial model. In Chapter 10, we will use this model to derive the famous Black–Scholes option pricing formula.

Theorem 6.4 *Let* $\mathbb{M}$ *be an arbitrage-free binomial model with lifetime* L *and with* T *time increments each of length* Δt. *Let* p *be the probability of an up-tick in the stock price from* S *to* Su *and let* $q = 1 - p$ *be the probability of a down-tick from* S *to* Sd. *Then the stock price is given by*

$$S_T = S_0 e^{\mu L + \sigma \sqrt{\Delta t} \sum_{i=1}^{T} X_i}$$

where the **drift** μ *and* **local volatility** σ *are defined by*

$$\mu = \frac{1}{\Delta t}(p \log u + q \log d)$$

$$\sigma = \frac{1}{\sqrt{\Delta t}}\sqrt{pq}(\log u - \log d)$$

The random walk portion of the stock price is given by

$$Y_T = \sum_{i=1}^{T} X_i$$

where the random variables X_i are independent standard Bernoulli random variables defined by

$$X_i = \begin{cases} \frac{q}{\sqrt{pq}} & \text{with probability } p \\ \frac{-p}{\sqrt{pq}} & \text{with probability } q \end{cases} \qquad \square$$

How to Choose the Tick Parameters

Returning to the issue of choosing the tick parameters u and d for a binomial model, let us refer to the following parameters as the **statistical parameters** of the model:

1) the drift (expected instantaneous return)

$$\mu = \frac{1}{\Delta t}(p \log u + q \log d)$$

2) the local volatility

$$\sigma = \frac{1}{\sqrt{\Delta t}}\sqrt{pq}(\log u - \log d)$$

3) The instantaneous expected return

$$\nu = \frac{1}{\Delta t}\log(pu + qd)$$

Assuming that we have estimated values for σ and one of μ or ν based on historical pricing data, we can use these formulas to get estimates of the tick parameters and the natural probability. The two most common ways to do this are described below.

The Cox–Ross–Rubinstein Binomial Model

Suppose we have estimated the stock's drift μ and volatility σ using historical pricing data. Let us call these estimates $\widehat{\mu}$ and $\widehat{\sigma}$, respectively. Then we determine the values of u, d and p in such a way that $\widehat{\mu}$ and μ are equal (or approximately equal) and similarly for σ and $\widehat{\sigma}$.

In 1979, Messieurs Cox, Ross and Rubinstein published a paper in which they set

$$u = e^{\widehat{\sigma}\sqrt{\Delta t}}, \quad d = u^{-1} = e^{-\widehat{\sigma}\sqrt{\Delta t}} \quad \text{and} \quad p = \frac{1}{2}\left[1 + \frac{\widehat{\mu}}{\widehat{\sigma}}\sqrt{\Delta t}\right]$$

Then

$$\begin{aligned}
\mu &= \frac{1}{\Delta t}[p\log u + (1-p)\log d] \\
&= \frac{1}{\Delta t}[p\log(u/d) + \log d] \\
&= \frac{1}{\Delta t}\left[2p\widehat{\sigma}\sqrt{\Delta t} - \widehat{\sigma}\sqrt{\Delta t}\right] \\
&= \frac{\widehat{\sigma}}{\sqrt{\Delta t}}[2p - 1] \\
&= \frac{\widehat{\sigma}}{\sqrt{\Delta t}}\frac{\widehat{\mu}}{\widehat{\sigma}}\sqrt{\Delta t} \\
&= \widehat{\mu}
\end{aligned}$$

and

$$\begin{aligned}
\sigma^2 &= \frac{1}{\Delta t}pq[\log(u/d)]^2 \\
&= \frac{1}{\Delta t}\frac{1}{4}\left[1 + \frac{\widehat{\mu}}{\widehat{\sigma}}\sqrt{\Delta t}\right]\left[1 - \frac{\widehat{\mu}}{\widehat{\sigma}}\sqrt{\Delta t}\right]\left[2\widehat{\sigma}\sqrt{\Delta t}\right]^2 \\
&= \widehat{\sigma}^2\left[1 - \left(\frac{\widehat{\mu}}{\widehat{\sigma}}\right)^2\Delta t\right] \\
&= \widehat{\sigma}^2 - \widehat{\mu}^2\Delta t
\end{aligned}$$

which approaches $\widehat{\sigma}^2$ as Δt approaches 0. The binomial model with these parameters u, d and p is called the **Cox–Ross–Rubinstein** **model** or **CRR model**. Note, however, that these formulas do not always make sense, since the condition $0 \le p \le 1$ implies that $|\widehat{\mu}/\widehat{\sigma}| \le 1$, which may not be the case.

The Equal-Probability Binomial Model

Another approach to choosing the parameters is to set $p = 1/2$ and set the instantaneous expected return ν and the volatility σ to estimated values $\widehat{\nu}$ and $\widehat{\sigma}$, respectively; that is,

$$\widehat{\sigma} = \frac{1}{2\sqrt{\Delta t}}(\log u - \log d)$$

and

$$\hat{\nu} = \frac{1}{\Delta t} \log\left(\frac{u+d}{2}\right)$$

Then

$$u + d = 2e^{\hat{\nu}\Delta t}$$

and

$$u/d = e^{2\hat{\sigma}\sqrt{\Delta t}}$$

and so

$$u = \frac{2e^{\hat{\nu}\Delta t + 2\hat{\sigma}\sqrt{\Delta t}}}{1 + e^{2\hat{\sigma}\sqrt{\Delta t}}}, \quad d = \frac{2e^{\hat{\nu}\Delta t}}{1 + e^{2\hat{\sigma}\sqrt{\Delta t}}}$$

The model with these parameters is called the **equal-probability binomial model**.

Exercises

1. A certain stock is currently selling for $50. The feeling is that for each month over the next 2 months, the stock's price will rise by 10% or fall by 10%. Assuming a risk-free rate of 1%, calculate the price of a European call with strike price K given by
    a) $52   b) $51   c) $50
    d) $49   e) $48   f) $47
    What about a European put with the same strike price and expiration date?

2. A certain stock is currently selling for $10. The feeling is that for each month over the next 2 months, the stock's price will rise by 5% or fall by 10%. Assuming a risk-free rate of 1%, calculate the price of a European call with strike price K given by
    a) $11   b) $10   c) $9   d) $8
    What about a European put with the same strike price and expiration date?

3. Referring to Example 6.1, explain why there is a *loss* in all states except the first; that is, there is a loss with probability $3/4$.

4. Show that $\{U, D\}^k$ has size 2^k. *Hint*: Use mathematical induction or the fundamental counting principle (also known as the multiplication rule).

5. Show that

$$X_i = \begin{cases} \frac{q}{\sqrt{pq}} & \text{with probability } p \\ \frac{-p}{\sqrt{pq}} & \text{with probability } q \end{cases}$$

6. Show that the two values of a Bernoulli random variable X with $p = 1/2$ are given by $\mathcal{E}(X) \pm \sqrt{\text{Var}(X)}$.

7. An alternative X that depends on the final state only through the number of U's in the state is called a **path-independent alternative**. In particular, if $\mathcal{P}$ is the partition of Ω whose blocks are the subsets G_k of Ω that contain exactly k U's

$$G_k = \{\omega \in \Omega \mid N_U(\omega) = k\}$$

then X is path-independent if and only if there are constants X_k for which

$$X_k = X(\text{any } \omega \in G_k)$$

for $k = 0, \ldots, T$.

a) Show that

$$|G_k| = \binom{T}{k}$$

b) Show that the probability (under the martingale measure) of any $\omega \in G_k$ is

$$\pi_U^{N_U(\omega)}(1 - \pi_U)^{T-N_U(\omega)} = \pi_U^k(1 - \pi_U)^{T-k}$$

c) Show that the probability of G_k is

$$\mathbb{P}_\Pi(G_k) = \binom{T}{k}\pi_U^k(1 - \pi_U)^{T-k}$$

d) Show that if X is a path-independent alternative, then

$$\mathcal{I}_0(X) = e^{-rL}\sum_{k=0}^{T} X_k \binom{T}{k}\pi_U^k(1 - \pi_U)^{T-k}$$

8. Write a computer program or an Excel spreadsheet to compute the price of a European call under the binomial model where $T = 2$.

9. Verify that

$$\mathcal{E}_p(\log E_i) = p\log u + q\log d$$
$$\text{Var}_p(\log E_i) = pq(\log u - \log d)^2$$

10. In a general discrete-time model, knowledge of the state of the economy at a given time implies knowledge of the asset prices at that time. Why? Is the converse necessarily true? What if at time t_k we know all previous states and asset prices? Support your answer. What happens in the case of the binomial model?

Chapter 7
Pricing Nonattainable Alternatives in an Incomplete Market

In this chapter, we discuss the problem of pricing *nonattainable* alternatives in an incomplete discrete-time model. It will be convenient in this chapter to fix an order $\Omega = (\omega_1, \ldots, \omega_m)$ for the state space and think of all random variables $X : \Omega \to \mathbb{R}$ as vectors

$$X = (X(\omega_1), \ldots, X(\omega_m))$$

In this chapter, we denote the discounted initial pricing functional $\overline{\mathcal{I}}_0$ simply by $\mathcal{I}$. The overbar will be used in its customary way not to denote discounting but to denote an extension of a function; thus, $\overline{f}$ is an extension of f.

Incompleteness in a Discrete-Time Model

It is often the case that a discrete model is incomplete. Consider, for example, a single-period model. In such a model, a trading strategy reduces to just a single portfolio $\Theta = (\theta_1, \ldots, \theta_n)$, acquired at time t_0 and held for the single period of the model. Hence, θ_i is a *constant* for each i.

In order for an alternative $X = (x_1, \ldots, x_m)$ to be attainable, there must be a portfolio Θ that satisfies the following system of equations:

$$S_{T,1}(\omega_1)\theta_1 + \cdots + S_{T,n}(\omega_1)\theta_n = x_1$$
$$\vdots$$
$$S_{T,1}(\omega_m)\theta_1 + \cdots + S_{T,n}(\omega_m)\theta_n = x_m$$

But if $m > n$, then there are more equations than variables and so there cannot be a solution Θ for all possible vectors $X = (x_1, \ldots, x_m)$. Hence, if the number of states in a discrete-time model exceeds the number of assets, the model is incomplete.

Mathematical Background

Before continuing, we wish to discuss some issues related to strongly positive vectors and linear functionals that will be needed later. We write $S \leq \mathbb{R}^m$ to denote the fact that S is a subspace of $\mathbb{R}^m$ and $S < \mathbb{R}^m$ to denote the fact that S is a *proper* subspace of $\mathbb{R}^m$.

Strongly Positive Vectors

We begin with a result on strongly positive vectors.

Theorem 7.1

1) *If $Z \in \mathbb{R}^m$ is strongly positive, then all vectors "close" to Z are also strongly positive; that is, there is a real number $\epsilon > 0$ such that all vectors $X \in S$ that satisfy $|X - Z| < \epsilon$ are also strongly positive.*
2) *If $S \leq \mathbb{R}^m$ and $Z \in S$ is strongly positive, then S has a basis consisting of strongly positive vectors.*

Proof. For part 1), write $Z = (z_1, \ldots, z_m)$ and let $\epsilon = \min\{z_i\}$. If $|X - Z| < \epsilon$ then

$$z_i - x_i \leq |z_i - x_i| \leq |Z - X| < \epsilon \leq z_i$$

and so we must have $x_i > 0$ for all i, whence X is strongly positive.

For part 2), let $\mathcal{B} = \{Z_1, Z_2, \ldots, Z_k\}$ be a basis for S, where $Z = Z_1$ and let

$$\mathcal{D} = \{Z_1, Z_1 + \lambda Z_2, \ldots, Z_1 + \lambda Z_k\}$$

where $\lambda > 0$. To see that $\mathcal{D}$ is linearly independent, if

$$a_1 Z_1 + a_2(Z_1 + \lambda Z_2) + \cdots + a_k(Z_1 + \lambda Z_k) = 0$$

then

$$\left(\sum a_i\right) Z_1 = -\lambda(a_2 Z_2 + \cdots + a_k Z_k)$$

and so $\sum a_i = 0$ and therefore $a_i = 0$ for all i. Thus, $\mathcal{D}$ is a basis for S. Now,

$$d_k = \|Z_1 - (Z_1 + \lambda Z_k)\| = \lambda \|Z_k\|$$

and so if λ is sufficiently small so are the distances d_k and part 1) implies that all vectors in $\mathcal{D}$ are strongly positive.$\square$

Linear Functionals and Their Kernels

We begin with a definition.

Definition *A* **linear functional** *on $S \leq \mathbb{R}^m$ is a function $f: S \to \mathbb{R}$ satisfying*

$$f(aX + bY) = af(X) + bf(Y)$$

for all $a, b \in \mathbb{R}$ and $X, Y \in S$. The set of all linear functionals on S is denoted *by S^* and called the* **dual space** *of S. The* **kernel** *of a linear functional f is the subspace*

$$\ker(f) = \{X \in S \mid f(X) = 0\} \qquad\qquad \square$$

A nonzero linear functional $f \in S^*$ is completely determined by its value on any vector $W \in S$ that is *not* in $\ker(f)$. To see this, we decompose S into a direct sum

$$S = \ker(f) \oplus \langle W \rangle$$

To verify this decomposition, note first that $\ker(f) \cap \langle W \rangle = \{0\}$, since if $X \in \ker(f) \cap \langle W \rangle$, then $X = \lambda W$ for some λ and so

$$0 = f(X) = \lambda f(W)$$

which implies that $\lambda = 0$, whence $X = 0$. Also, for any $X \in S$,

$$X = \left(X - \frac{f(X)}{f(W)} W \right) + \frac{f(X)}{f(W)} W \in \ker(f) + \langle W \rangle$$

and so $S = \ker(f) \oplus \langle W \rangle$. Thus, any $X \in S$ can be written in the form

$$X = Z + aW$$

where $Z \in \ker(f)$ and $a \in \mathbb{R}$ and so

$$f(X) = f(Z + aW) = f(Z) + af(W) = af(W)$$

It follows that f is completely determined by its value on any $W \in S \setminus \ker(f)$.

As a consequence, if $f, g \in S^*$ have the same kernel K, then f and g are scalar multiples of one another; that is, $g = \lambda f$ for some nonzero $\lambda \in \mathbb{R}$. This is clear if f or g is zero. If not, let

$$S = K \oplus \langle W \rangle$$

where $W \in S \setminus K$. Then for any $X = Z + aW$ with $Z \in K$, we have

$$\lambda f(X) = \lambda a f(W) = ag(W) = g(X)$$

and so $g = \lambda f$.

Theorem 7.2 *Let $S \leq \mathbb{R}^m$ and let $f, g \in S^*$.*
1) If $W \in S \setminus \ker(f)$, then

$$S = \ker(f) \oplus \langle W \rangle$$

2) If $\ker(f) = \ker(g)$, then there is a real number λ for which $g = f\lambda$. $\square$

The Representation Theorem for Linear Functionals

If $\mathcal{S} \leq \mathbb{R}^m$ and if $Y \in \mathcal{S}$, then the inner product by Y defines a linear functional on $\mathcal{S}$. Specifically, the function on $\mathcal{S}$ defined by

$$I_Y(X) = \langle X, Y \rangle$$

defines a linear functional on $\mathcal{S}$, called the **inner product by Y**.

The following important theorem says that all linear functionals are given in this way. We will denote the standard basis vectors in $\mathbb{R}^m$ by $E_1, \ldots, E_m$; that is,

$$E_k = (0, \ldots, 0, 1, 0, \ldots, 0)$$

where the 1 is in the kth position.

Theorem 7.3 (The Riesz Representation Theorem) *Let $\mathbb{R}^m$ have the standard inner product.*
1) *If $\mathcal{S} \leq \mathbb{R}^m$ and $f \in \mathcal{S}^*$, then there is a unique vector $Y_f \in \mathcal{S}$ such that*

$$f(X) = \langle X, Y_f \rangle = I_{Y_f}$$

for all $X \in \mathcal{S}$; that is,

$$f = I_{Y_f}$$

*We refer to Y_f as the **Riesz vector** for f.*
2) *Thus, the map $\mathcal{R}: \mathcal{S}^* \to \mathcal{S}$ defined by*

$$\mathcal{R}(f) = Y_f$$

is a bijection, whose inverse is

$$\mathcal{R}^{-1}(Y) = I_Y$$

3) *The Riesz vector for a linear functional f on $\mathbb{R}^m$ is*

$$Y_f = (f(E_1), \ldots, f(E_m))$$

Proof. For part 1), if $f = 0$ then we can take $Y_f = 0$. If $f \neq 0$, then let $Z \in (\ker(f))^{\perp}$ and write

$$\mathcal{S} = \ker(f) \oplus \langle Z \rangle$$

Now, $f = I_W$ for some vector W if and only if

$$f = I_W \text{ on } \ker(f) \quad \text{and} \quad f(Z) = I_W(Z)$$

that is, if and only if

$$W \in (\ker(f))^{\perp} \quad \text{and} \quad f(Z) = \langle Z, W \rangle$$

But since $(\ker(f))^{\perp}$ is one-dimensional, the first condition holds if and only if

$W = \lambda Z$ for some scalar λ. In this case, the second condition holds if and only if $f(Z) = \langle Z, \lambda Z \rangle$; that is, if and only if

$$\lambda = \frac{f(Z)}{\langle Z, Z \rangle}$$

Hence,

$$Y_f = \frac{f(Z)}{\langle Z, Z \rangle} Z$$

for any $Z \in (\ker(f))^{\perp}$. We leave proof of part 2) and part 3) to the reader.$\square$

Extensions of Linear Functionals

In general, a linear functional on a proper subspace of $\mathbb{R}^m$ has many extensions to $\mathbb{R}^m$. Here is the formal definition of an extension.

Definition *Let $S \leq \mathbb{R}^m$ and let $f \in S^*$. An **extension** of f to $\mathbb{R}^m$ is a linear functional $\overline{f}$ on $\mathbb{R}^m$ for which*

$$\overline{f}(X) = f(X)$$

for all $X \in S$.$\square$

It is a well known fact from linear algebra that if $S \leq \mathbb{R}^m$, then a linear functional f on S can be extended to a linear functional $\overline{f}$ on $\mathbb{R}^m$ as follows. Extend a basis $\{X_1, \ldots, X_k\}$ for S to a basis

$$\{X_1, \ldots, X_k, Y_1, \ldots, Y_{m-k}\}$$

for $\mathbb{R}^m$. Then set $\overline{f}(X_i) = f(X_i)$ for all i and define $\overline{f}(Y_j)$ arbitrarily for all j. The function $\overline{f}$ defines a unique linear functional on $\mathbb{R}^m$ that extends f.

The following theorem characterizes extensions of a linear functional in terms of their Riesz vectors. By way of notation, if $X \in \mathbb{R}^m$ and $S \leq \mathbb{R}^m$, then we set

$$X + S = \{X + Y \mid Y \in S\}$$

Theorem 7.4 *Let $S \leq \mathbb{R}^m$ and let $f \in S^*$. Let $K = \ker(f)$.*
1) *A linear functional $g: \mathbb{R}^m \to \mathbb{R}$ is an extension of f if and only if*

$$Y_g \in Y_f + S^{\perp}$$

2) *Let $f \neq 0$. For each $U \in K^{\perp}$ that is not orthogonal to Y_f; that is, for each $U \in K^{\perp} \setminus \{Y_f\}^{\perp}$, there is a unique scalar*

$$\lambda = \frac{\langle Y_f, Y_f \rangle}{\langle Y_f, U \rangle}$$

for which $I_{\lambda U}$ is an extension of f. Moreover, all extensions of f have this form.

Proof. For part 1), $f = g$ on $\mathcal{S}$ if and only if $f - g$ is 0 on $\mathcal{S}$; that is, if and only if $Y_g - Y_f = Y_{f-g} \in \mathcal{S}^\perp$. For part 2), since

$$\mathcal{S} = K \oplus \langle Y_f \rangle$$

a linear functional I_U on $\mathbb{R}^m$ is an extension of f if and only if

$$I_U(K) = \{0\} \quad \text{and} \quad I_U(Y_f) = f(Y_f)$$

that is, if and only if

$$U \in K^\perp \quad \text{and} \quad \langle Y_f, U \rangle = \langle Y_f, Y_f \rangle$$

Since $f \neq 0$, the second condition requires that $\langle Y_f, U \rangle \neq 0$; that is, $U \notin \{Y_f\}^\perp$.

Now, if a vector $U \in K^\perp \setminus \{Y_f\}^\perp$ does not satisfy the second condition, some scalar multiple λU of U will, since the second condition

$$\langle Y_f, \lambda U \rangle = \langle Y_f, Y_f \rangle$$

can be solved for a unique λ, namely,

$$\lambda = \frac{\langle Y_f, Y_f \rangle}{\langle Y_f, U \rangle}$$

Thus, for each $U \in K^\perp \setminus \{Y_f\}^\perp$, there is a unique scalar λ for which $I_{\lambda U}$ is an extension of f and this accounts for all extensions of f.$\square$

Positive Linear Functionals

We define positivity for linear functionals in a manner consistent with our definition of positivity for vectors.

Definition *Let $f \colon \mathcal{S} \to \mathbb{R}$ be a linear functional on $\mathcal{S} \subseteq \mathbb{R}^m$. Then*
1) *f is **nonnegative**, written $f \geq 0$, if*

$$X > 0 \quad \Rightarrow \quad f(X) \geq 0$$

for all $X \in \mathcal{S}$.
2) *f is **strictly positive**, written $f > 0$, if f is nonnegative and nonzero.*
3) *f is **strongly positive**, written $f \gg 0$, if*

$$X > 0 \quad \Rightarrow \quad f(X) > 0$$

for all $X \in \mathcal{S}$.$\square$

It is not hard to see that in a model without arbitrage the initial pricing functional $\mathcal{I}$ is strongly positive, for if $X > 0$ is an attainable alternative whose initial price $\mathcal{I}(X)$ is not positive, then there is arbitrage.

The following theorem characterizes positivity for linear functionals on $\mathbb{R}^m$ in terms of Riesz vectors. We leave the proof to the reader.

Theorem 7.5 *Let* $f: \mathbb{R}^m \to \mathbb{R}$ *be a nonzero linear functional. Then*
1) *f is strictly positive if and only if Y_f is strictly positive.*
2) *f is strongly positive if and only if Y_f is strongly positive.* $\square$

It is worth pointing out that this theorem applies only to linear functionals defined on all of $\mathbb{R}^m$. For instance, consider the subspace

$$S = \{(a, -a) \mid a \in \mathbb{R}\}$$

of $\mathbb{R}^2$. Since there are no strictly positive vectors in S, it follows that all linear functionals on S are strongly positive, even though the Riesz vector is not strongly positive.

Extensions and Positivity

Let us consider the issue of whether a strongly positive linear functional on a subspace S can be extended to a strongly positive functional on $\mathbb{R}^m$. For any $f \in S^*$, let

$$\mathcal{E}_{>0}(f) = \{\text{strictly positive extensions of } f\}$$
$$\mathcal{E}_{\gg 0}(f) = \{\text{strongly positive extensions of } f\}$$

Of course,

$$\mathcal{E}_{\gg 0}(f) \subseteq \mathcal{E}_{>0}(f)$$

Theorem 7.6 *Let* $S \le \mathbb{R}^m$. *Then every strongly positive linear functional f on S has a strongly positive extension to $\mathbb{R}^m$. Hence, $\mathcal{E}_{\gg 0}(f)$ is nonempty.*
Proof. Let $K = \ker(f)$. If there is a strongly positive vector $W \in S^\perp$, then we may choose a $\lambda > 0$ large enough so that $Y_f + \lambda W \in Y_f + S^\perp$ is strongly positive. Then the linear functional $I_{Y_f + \lambda W}$ is a strongly positive extension of f. Thus, we may assume that $S^\perp$ does not contain a strongly positive vector, in which case Theorem A.8 of Appendix A implies that $(S^\perp)^\perp = S$ contains a strictly positive vector.

Now, since f is strongly positive, K contains no strictly positive vectors and so Theorem A.7 implies that $K^\perp$ contains a strongly positive vector W. Furthermore, Theorem 7.1 implies that $K^\perp$ has a basis $\mathcal{B} = \{W_1, \ldots, W_k\}$ consisting entirely of strongly positive vectors.

But Y_f cannot be orthogonal to all of the basis vectors W_i, since then Y_f would be orthogonal to $K^\perp$; that is, Y_f would be in K, which is not the case. Hence, we may assume that $W_i \in K^\perp \setminus \{Y_f\}^\perp$ for some i and so Theorem 7.4 implies that there is a unique scalar λ for which $I_{\lambda W_i}$ is an extension of f.

To show that $I_{\lambda W_i}$ is strongly positive, we use the strictly positive vector X in $\mathcal{S}$. In particular,

$$\lambda\langle X, W_i\rangle = \langle X, \lambda W_i\rangle = I_{\lambda W_i}(X) = f(X) > 0$$

and since $\langle X, W_i\rangle > 0$, it follows that $\lambda > 0$. Thus, λW_i is strongly positive and therefore $I_{\lambda W_i}$ is a strongly positive extension of f.$\square$

Pricing Nonattainable Alternatives

We can now discuss the issue at hand, namely pricing a nonattainable alternative $W \notin \mathcal{M}$ in an incomplete discrete-time model $\mathbb{M}$. (Recall that $\mathcal{M}$ is the subspace of attainable alternatives.) Since the replicating alternative pricing procedure cannot be used to assign a fair value to W, we must find another way to price W.

If $X = (x_1, \ldots, x_m)$ and $Y = (y_1, \ldots, y_m)$ are vectors in $\mathbb{R}^m$, we say that X **dominates** Y, written $X \geq Y$, if $X - Y \geq 0$; that is, if $x_k \geq y_k$ for all k.

Minimum Dominating Price

Let us consider the question from the point of view of an investor who wants to sell the payoff W. The investor knows that he cannot *duplicate* the payoff with a self-financing trading strategy. However, in order to hedge the risk of the short position, the investor can use any self-financing trading strategy Φ whose payoff *dominates* W; that is, for which

$$\mathcal{V}_T(\Phi) \geq W$$

Put another way, if

$$\mathcal{D}_W(\mathcal{M}) = \{X \in \mathcal{M} \mid X \geq W\}$$

is the set of all *attainable* alternatives that dominate W, then the investor can hedge W by investing in any self-financing trading strategy that replicates any vector in $\mathcal{D}_W(\mathcal{M})$. Thus, at least from the seller's point of view, it seems reasonable to set the fair price of W to be the *minimum price* of the members of $\mathcal{D}_W(\mathcal{M})$.

Definition *If W is a nonattainable alternative, then the* **minimum dominating price** *of W is*

$$P^d(W) = \min_{X \in \mathcal{D}_W(\mathcal{M})} \mathcal{I}(X) \qquad \square$$

Of course, this definition only makes sense if $\mathcal{D}_W(\mathcal{M})$ is nonempty, so let us address this issue forthwith.

Theorem 7.7 *The set $\mathcal{M}$ of all attainable vectors for a discrete-time model contains a strongly positive vector. Moreover, $\mathcal{D}_W(\mathcal{M})$ is nonempty for all $W \in \mathbb{R}^m$.*

Proof. Let Φ be the trading strategy that invests $\alpha > 0$ units in the risk-free asset and continually rolls it over. Then $\mathcal{V}_T(\Phi) \in \mathcal{M}$ and $\mathcal{V}_T(\Phi)(\omega) \geq \alpha > 0$ for all $\omega \in \Omega$. For the second statement, if $W = (w_1, \ldots, w_m)$ then we need only choose α so that $\alpha \geq w_i$ for all i.$\square$

We prove in Appendix B that the minimum dominating price of any nonattainable alternative W is actually achieved by some dominating attainable alternative; that is, there is an $X_W \in \mathcal{D}_W(\mathcal{M})$ for which

$$\mathcal{I}(X_W) = P^d(W)$$

Maximum Extension Price

There is another viewpoint that we can take when trying to price a nonattainable alternative $W \notin \mathcal{M}$. Namely, the initial pricing functional $\mathcal{I}$ has domain $\mathcal{M}$ and we can extend it to a linear functional $\overline{\mathcal{I}}$ on $\mathbb{R}^m$, perhaps in infinitely many ways. What about pricing W as the price $\overline{\mathcal{I}}(W)$ of one of these extensions? In fact, since the initial pricing functional $\mathcal{I}$ is strongly positive, perhaps we should restrict attention only to strongly positive extensions of $\mathcal{I}$. Theorem 7.6 shows that the set $\mathcal{E}_{\gg 0}(\mathcal{I})$ is nonempty and so the following definition makes sense.

Definition *For a nonattainable alternative W, the* **maximum extension price** *is*

$$P^e(W) = \max\{\overline{\mathcal{I}}(W) \mid \overline{\mathcal{I}} \in \mathcal{E}_{\gg 0}(\mathcal{I})\} \qquad\qquad \square$$

In the exercises, we ask the reader to show that it doesn't matter whether we restrict attention to nonnegative, strictly positive or strongly positive extensions since

$$\max\{\overline{\mathcal{I}}(W) \mid \overline{\mathcal{I}} \in \mathcal{E}_{\geq 0}(\mathcal{I})\} = \max\{\overline{\mathcal{I}}(W) \mid \overline{\mathcal{I}} \in \mathcal{E}_{> 0}(\mathcal{I})\}$$
$$= \max\{\overline{\mathcal{I}}(W) \mid \overline{\mathcal{I}} \in \mathcal{E}_{\gg 0}(\mathcal{I})\}$$

for all $W \notin \mathcal{M}$.

We also prove in Appendix B that the maximum extension price is always achieved by some nonnegative extension; that is, there is a nonnegative extension $\overline{\mathcal{I}}$ of $\mathcal{I}$ for which

$$\overline{\mathcal{I}}(W) = P^e(W)$$

However, the following example shows that there need not be a *strongly* positive extension that achieves the maximum extension price.

Example 7.1 Suppose that $\mathcal{M} = \{(x, x) \mid x \in \mathbb{R}\} \subseteq \mathbb{R}^2$ and let $\mathcal{I}((x, x)) = ax$ for some $a > 0$. Then

$$\mathcal{I}((x, x)) = \langle (x, x), (a/2, a/2) \rangle$$

and so $Y_{\mathcal{I}} = (a/2, a/2)$. Any extension $\overline{\mathcal{I}}$ of $\mathcal{I}$ has the form

$$\overline{\mathcal{I}}(X) = \langle X, Y_{\mathcal{I}} + W \rangle$$

where $W \in \mathcal{M}^{\perp}$. Hence, W has the form $(z/2, -z/2)$ for $z \in \mathbb{R}$ and we have

$$Y_{\overline{\mathcal{I}}} = \left(\frac{a}{2}, \frac{a}{2} \right) + \left(\frac{z}{2}, -\frac{z}{2} \right) = \frac{1}{2}(a + z, a - z)$$

Thus, $\overline{\mathcal{I}}$ is strictly positive if and only if $-a \le z \le a$ and $\overline{\mathcal{I}}$ is strongly positive if and only if $-a < z < a$. Now let $W = (0, 1)$, which is not attainable. Then

$$\overline{\mathcal{I}}(W) = \langle W, Y_{\overline{\mathcal{I}}} \rangle = \frac{1}{2}(a - z)$$

and so

$$P^e(W) = \max\{\overline{\mathcal{I}}(W) \mid \overline{\mathcal{I}} \in \mathcal{E}_{>0}(\mathcal{I})\} = \max_{-a \le z \le a} \left\{ \frac{1}{2}(a - z) \right\} = a$$

But no strongly positive extension achieves this maximum. In fact, the only extension that achieves this maximum is the one for which $z = -a$, and this extension is only strictly positive.$\square$

Optimal Solutions to the Pricing Problem

Thus, for a nonattainable alternative W we have two prices: the minimum dominating price

$$P^d(W) = \min\{\mathcal{I}(X) \mid X \in \mathcal{D}_W\}$$

and the maximum extension price

$$P^e(W) = \max\{\overline{\mathcal{I}}(W) \mid \overline{\mathcal{I}} \in \mathcal{E}_{\gg 0}(\mathcal{I})\}$$

and we have said that both these prices are achieved. Moreover, it is a pleasant fact that the two prices are the same; that is,

$$P^d(W) = P^e(W)$$

We also prove this in Appendix B. However, in the exercises, we ask the reader to prove that

$$-\infty < P^e(W) \le P^d(W) < \infty$$

The following theorem summarizes our results.

Theorem 7.8 *Let W be a nonattainable alternative.*

1) *The sets $\mathcal{E}_{\gg 0}(\mathcal{I})$ and $\mathcal{D}_W$ are nonempty and*

$$P^d(W) = P^e(W)$$

that is,

$$\min\{\mathcal{I}(X) \mid X \in \mathcal{D}_W\} = \max\{\overline{\mathcal{I}}(W) \mid \overline{\mathcal{I}} \in \mathcal{E}_{\gg 0}(\mathcal{I})\}$$

2) *There is a dominating attainable alternative $X_W \in \mathcal{M}$ for which*

$$\mathcal{I}(X_W) = P^d(W)$$

3) *There is a nonnegative extension $\overline{\mathcal{I}}_W$ of $\mathcal{I}$ for which*

$$\overline{\mathcal{I}}_W(W) = P^e(W) \qquad\qquad \square$$

We can now make the following definition.

Definition *The* **fair price** *of a nonattainable alternative W is defined to be*

$$\max\{\overline{\mathcal{I}}(W) \mid \overline{\mathcal{I}} \in \mathcal{E}_{\gg 0}(\mathcal{I})\} = \min\{\mathcal{I}(X) \mid X \in \mathcal{D}_W\} \qquad \square$$

Exercises

1. Suppose that f is a linear functional on $\mathcal{S}$ for which

$$x \gg 0 \quad \Rightarrow \quad f(x) > 0$$

for all $x \in \mathcal{S}$. Show that f need not be strongly positive.

2. If $f : \mathbb{R}^m \to \mathbb{R}$ is a linear functional on $\mathbb{R}^m$ prove that

$$Y_f = \langle f(E_1), \ldots, f(E_m) \rangle$$

where E_i are the standard basis vectors for $\mathbb{R}^m$.

3. If f is a linear functional on $\mathbb{R}^m$, prove the following:
    a) f is strictly positive if and only if Y_f is strictly positive.
    b) f is strongly positive if and only if Y_f is strongly positive.

4. For Example 7.1, verify that

$$\max\{\overline{\mathcal{I}}(W) \mid \overline{\mathcal{I}} \in \mathcal{E}_{>0}(\mathcal{I})\} = \min_{X \in \mathcal{D}_W} \mathcal{I}(X)$$

5. Let $\mathcal{M} = \{(2a, 3a) \mid a \in \mathbb{R}\} \subseteq \mathbb{R}^2$ and let $\mathcal{I}(2, 3) = \alpha$. Show that the price of an arbitrary alternative $W = (x, y)$ is

$$\max\left\{ \frac{\alpha x}{2}, \frac{\alpha y}{3} \right\}$$

6. Let S be a subspace of $\mathbb{R}^n$ and suppose that S contains a strongly positive vector. Prove that any vector in S can be written as the difference of two nonnegative vectors in S.

7. Let W be a nonattainable alternative. Prove that

$$-\infty < \max\{\overline{\mathcal{I}}(W) \mid \overline{\mathcal{I}} \in \mathcal{E}_{>0}(\mathcal{I})\} \le \min\{\mathcal{I}(X) \mid X \in \mathcal{D}_W\} < \infty$$

8. Suppose that f is a strictly positive linear functional on $\mathbb{R}^n$ and that $X \in \mathbb{R}^n$. Prove that for any $\epsilon > 0$, we can find a strongly positive linear functional g for which

$$|g(X) - f(X)| < \epsilon$$

Chapter 8
Optimal Stopping and American Options

The models that we have created thus far are designed to price European options, which can only be exercised at the expiration time. However, in the real world, most stock options are of the American variety. In this chapter, we want to take a look at the issue of pricing American options, which can be exercised at any time between the purchase date and the expiration date.

We can give the gist of the upcoming discussion quite simply as follows. For an American option the final payoff, which is a random variable on the state space Ω, depends upon the strategy that the investor uses in determining when to exercise the option. In symbols, for each exercise strategy τ (to be made more precise later), there is a final payoff random variable $\mathbb{X}_\tau \colon \Omega \to \mathbb{R}$.

Now, it would seem prudent to choose an exercise strategy τ that maximizes the final payoff $\mathbb{X}_\tau$. However, the payoffs $\mathbb{X}_\tau$ are *functions* and so it is extremely unlikely that there is a single exercise strategy that will maximize the payoff for *all* states $\omega \in \Omega$.

On the other hand, if a final payoff $\mathbb{X}_\tau$ is attainable, then it has an *arbitrage-free initial price*, namely, the expected discounted payoff $\mathcal{E}_\Pi(\overline{X}_\tau)$ under the martingale measure Π. But these initial prices are *constants* and so an investor can choose an exercise strategy that maximizes these initial prices; that is, an investor can choose the *most valuable* exercise strategy. Any such strategy is called an *optimal exercise strategy*.

An Example

Let us set up a simple example to which we will refer later.

Example 8.1 Figure 8.1 shows the state tree for a binomial model whose nodes are labeled with the label S/P, where S is the stock price and $P = (x - 21)^+$ is the payoff for an American call C with strike price 21.

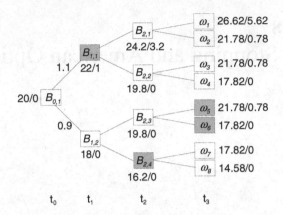

Figure 8.1: A binomial model state tree

We will assume that the risk-free rate r is 0. Note that for this model

$$T = 3,\ u = 1.1,\ d = 0.9$$

and the martingale measure up-tick probability is

$$\pi = \frac{1 - d}{u - d} = \frac{0.1}{0.2} = \frac{1}{2} \qquad \square$$

The Model

In general, our context will be a complete discrete-time model $\mathbb{M}$ with information structure

$$\mathbb{F} = \{\mathcal{P}_0, \dots, \mathcal{P}_T\}$$

The model is assumed to be arbitrage-free, with martingale measure Π, as is the case for Example 8.1. We also consider an American option, which can be exercised at any of the model's times

$$t_0 < t_1 < \cdots < t_T$$

We will assume a constant risk-free rate of r throughout the model.

The Payoff Process

For pricing European options, we have treated the final payoff of the option simply as an attainable alternative $Y \colon \Omega \to \mathbb{R}$. However, for pricing American options, we must consider an entire process $\mathbb{Y} = (Y_0, \dots, Y_T)$ of payoffs, where Y_k is the time-t_k return from the option if it is exercised at that time. We assume that $\mathbb{Y}$ is adapted to the filtration $\mathbb{F} = (\mathcal{P}_0, \dots, \mathcal{P}_T)$.

For our example, as shown in Figure 8.1, the payoffs of the American call C are

$$Y_3 = \max\{S_3 - 21, 0\} = \begin{cases} 5.62 & \text{for } \omega = \omega_1 \\ 0.78 & \text{for } \omega = \omega_2, \omega_3, \omega_5 \\ 0 & \text{otherwise} \end{cases}$$

$$Y_2 = \max\{S_2 - 21, 0\} = \begin{cases} 3.2 & \text{for } \omega \in B_{2,1} \\ 0 & \text{otherwise} \end{cases}$$

$$Y_1 = \max\{S_1 - 21, 0\} = \begin{cases} 1 & \text{for } \omega \in B_{1,1} \\ 0 & \text{otherwise} \end{cases}$$

and

$$Y_0 = 0$$

However, in keeping with the spirit of self-financing trading strategies, we will assume that the investor does not remove the funds from the model until the final time t_T and so any payoff is allowed to grow at the risk-free rate until the end of the model. Thus, if the investor decides to exercise at time t_k under state $\omega \in \Omega$, the time-t_k payoff (delivered at time t_T) will be

$$X_k(\omega) = e^{r(t_T - t_k)} Y_k(\omega)$$

This quantity is a bit easier to deal with, since all payoffs are now expressed in terms of a single date.

We refer to this (delayed) payoff process, or indeed to any stochastic process $\mathbb{X} = (X_0, \ldots, X_T)$ simply as a **payoff process** or as an **alternative process**. Our goal in this chapter is to decide when to exercise an alternative process.

Stopping Times

As shown in Figure 8.2, the decision about when to exercise an American option (or any payoff process) can be modeled as a random variable τ, called a **stopping time**, that identifies, for each time t_k, the outcomes $\omega \in \Omega$ for which the option should be exercised at that time; that is, $\tau(\omega) = k$ means exercise at time t_k if the final state is ω.

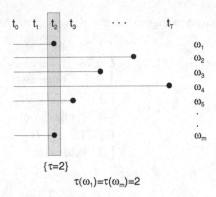

Figure 8.2: A stopping time

For $k < T$, the set $\{\tau = k\}$ of outcomes $\omega \in \Omega$ that have exercise time t_k is called the **stopping event** for time t_k. For the final time t_T, however, the stopping event consists of the outcomes for which we either exercise at time t_T or let the option expire worthless without exercising at all.

Of course, a stopping time is not much use if we cannot tell at a given time t_k whether or not to stop; that is, we must know the stopping event $\{\tau = k\}$ at time t_k. Put more mathematically, $\{\tau = k\}$ must be either empty or else a union of one or more blocks of the time-t_k state partition $\mathcal{P}_k$. Here is a formal definition.

Definition *Let $\mathcal{P}$ be a partition of Ω. We say that a subset $S \subseteq \Omega$ is $\mathcal{P}$-measurable if S is either empty or is the union of blocks of $\mathcal{P}$.* $\square$

Definition *A stopping time for $\mathbb{M}$ is a random variable*

$$\tau \colon \Omega \to \{0, \ldots, T\}$$

with the property that for all $k = 0, \ldots, T$, the stopping event

$$\{\tau = k\} = \{\omega \in \Omega \mid \tau(\omega) = k\}$$

is $\mathcal{P}_k$-measurable, where $\mathcal{P}_k$ is the time-t_k state partition. We denote the set of all stopping times with range $\{k, \ldots, k+j\}$ by $\mathcal{S}_{k,k+j}$. These are the stopping times that stop no earlier than time t_k and no later than time t_{k+j}. $\square$

Let us consider an important example of stopping times.

Example 8.2 It would be reasonable for an investor to tell his broker to sell a stock if the price reaches P or more dollars. This rule is a stopping time, referred to as the **first entry time** of the stock price process (S_k) into the interval $[P, \infty)$. Formally, it is defined as follows:

$$\tau(\omega) = \begin{cases} \min\{k \mid S_k(\omega) \geq P\} & \text{if } \{k \mid S_k(\omega) \geq P\} \neq \emptyset \\ T & \text{otherwise} \end{cases}$$

It is not hard to show that this is indeed a stopping time. Indeed, for $k < T$, the set $\{\tau = k\}$ is the set of all $\omega \in \Omega$ for which the price of the stock is less than P until time t_k, at which time the stock price is at least P. In symbols, if $k < T$ then

$$\{\tau = k\} = \{S_0 < P\} \cap \cdots \cap \{S_{k-1} < P\} \cap \{S_k \geq P\}$$

But each of the sets on the right is $\mathcal{P}_k$-measurable and therefore so is $\{\tau = k\}$. Similarly, for $k = T$, the set

$$\{\tau = T\} = \{S_0 < P\} \cap \cdots \cap \{S_{T-1} < P\}$$

is also $\mathcal{P}_k$-measurable and so τ is a stopping time.

It is also possible to show that the first entry time into any set that is the finite union of intervals (a, b), $[a, b)$, $(a, b]$, $[a, b]$ and/or rays $(-\infty, b)$, $(-\infty, b]$, $(a, -\infty)$ and $[a, -\infty)$ is also a stopping time. (This is true for any Borel set, defined in Chapter 9.) For example, the set

$$B = (-\infty, 17) \cup [20, \infty)$$

corresponds to the first time that the stock price drops below 17 or reaches 20. The shaded blocks in Figure 8.1 show the stopping events for the first entry time into B.□

Piecing Together Stopping Times

Under certain important situations, it is possible to piece together stopping times and get another stopping time.

Theorem 8.1 *Suppose that the time-t_k state partition is $\mathcal{P}_k = \{B_{k,1}, \ldots, B_{k,c}\}$ and that $\tau_u \in \mathcal{S}_{k,T}$ for each $u = 1, \ldots, c$. Then the function*

$$\tau^* = \sum_{u=1}^{c} \tau_u 1_{B_{k,u}}$$

is also a stopping time in $\mathcal{S}_{k,T}$.

Proof. Since $\tau_u \geq k$ for all u, it follows that $\tau^* \geq k$. Moreover, since $\mathcal{P}_k$ is a partition of Ω, for any $h \geq k$ we have

$$\{\tau^* = h\} = \bigcup_{u=1}^{c} (\{\tau^* = h\} \cap B_{k,u}) = \bigcup_{u=1}^{c} (\{\tau_u = h\} \cap B_{k,u})$$

But $B_{k,u}$ and $\{\tau_u = h\}$ are both unions of blocks of $\mathcal{P}_h$ and therefore so is their intersection. Hence, $\tau^* \in \mathcal{S}_{k,T}$.□

Payoff under a Stopping Time

A stopping time is a rule that determines when to take a payoff; that is, when to exercise. Indeed, applying a stopping time $\tau \in \mathcal{S}_{0,T}$ turns a payoff process $\mathbb{X}$ into a single payoff random variable $\mathbb{X}_\tau$ defined by

$$\mathbb{X}_\tau(\omega) = X_{\tau(\omega)}(\omega)$$

The random variable $\mathbb{X}_\tau$ is referred to as the **payoff** of the process $\mathbb{X}$ *under* the stopping time τ. We use the notation $\mathbb{X}_\tau$ rather than X_τ to emphasize the fact that the final payoff depends on the *entire* payoff stochastic process $\mathbb{X}$.

Example 8.3 Referring to Example 8.1, consider the stopping time τ shown in gray in Figure 8.1, which is the first entry time into

$$B = (-\infty, 17) \cup (20, \infty)$$

The payoff $\mathbb{X}_\tau$ is

$$\mathbb{X}_\tau(\omega) = \begin{cases} 1 & \text{if } \omega \in \{\omega_1, \omega_2, \omega_3, \omega_4\} \\ 0.78 & \text{if } \omega = \omega_5 \\ 0 & \text{if } \omega \in \{\omega_6, \omega_7, \omega_8\} \end{cases} \qquad \square$$

At any given time t_k, it seems logical that the investor should choose a stopping time $\tau \in \mathcal{S}_{k,T}$ that maximizes the payoff $\mathbb{X}_\tau$, among all members of $\mathcal{S}_{k,T}$. However, the payoffs $\mathbb{X}_\tau$ are *functions* on the state space Ω and it is extremely unlikely that there is a single stopping time τ^* that gives a better payoff for *all* states of the economy. For example, the stopping time σ that is defined by waiting until the end of the model in all cases (effectively turning the American option into a European option) is better than the stopping time τ in Example 8.3 for state ω_1 but not for state ω_2. Hence, we cannot say that either σ or τ is better than the other.

On the other hand, if $\mathbb{X}$ is a payoff process, then $\mathbb{X}_\tau$ has an arbitrage-free time-t_k price

$$\mathcal{I}_k(\mathbb{X}_\tau) = \mathcal{E}_\Pi(e^{-r(t_T - t_k)}\mathbb{X}_\tau \mid \mathcal{P}_k) = e^{-r(t_T - t_k)}\mathcal{E}_\Pi(\mathbb{X}_\tau \mid \mathcal{P}_k)$$

and we will show in a moment that there is a stopping time $\tau^* \in \mathcal{S}_{k,T}$ that maximizes this price among all stopping times in $\mathcal{S}_{k,T}$. Put somewhat colloquially, even though we cannot choose a stopping time that maximizes the payoff, we can choose a stopping time whose associated payoff is the most expensive!

Of course, a stopping time maximizes $\mathcal{I}_k(\mathbb{X}_\tau)$ if and only if it maximizes $\mathcal{E}_\Pi(\mathbb{X}_\tau \mid \mathcal{P}_k)$ and so we will work with the latter value. Accordingly, we make the following definitions.

Definition Let $\mathbb{X} = (X_0, \ldots, X_T)$ be an alternative process. Let Π be a martingale measure for the model $\mathbb{M}$. For any stopping time τ, let $\mathbb{X}_\tau$ be defined by

$$\mathbb{X}_\tau(\omega) = X_{\tau(\omega)}(\omega)$$

for $\omega \in \Omega$.

1) The **optimal value** of $\mathbb{X}$ for the interval $[k, T]$ is

$$\mathcal{M}_k(\mathbb{X}) = \max\{\mathcal{E}_\Pi(\mathbb{X}_\tau \mid \mathcal{P}_k) \mid \tau \in \mathcal{S}_{k,T}\}$$

The optimal value process $\mathbb{S} = (\mathcal{M}_0(\mathbb{X}), \ldots, \mathcal{M}_T(\mathbb{X}))$ is called the **Snell envelope** of $\mathbb{X}$. For simplicity, we will abbreviate $\mathcal{M}_k(\mathbb{X})$ by $\mathcal{M}_k$ when the process $\mathbb{X}$ is clear from the context.

2) An **optimal stopping time** for $\mathbb{X}$ over the interval $[k, T]$ is a stopping time τ^* that achieves the optimal value of $\mathbb{X}$; that is, for which

$$\mathcal{E}_\Pi(\mathbb{X}_{\tau^*} \mid \mathcal{P}_k) = \mathcal{M}_k(\mathbb{X}) \qquad \square$$

We should also remark that optimal stopping times represent the best *guess* as to when to stop based on *current* value, which is the best we can do without being able to see into the future. Thus, an optimal stopping time is not *guaranteed* to produce the best possible payoff and we will see an example later in which an optimal stopping time does fall short in this respect.

Example 8.4 Again referring to Example 8.1, we have seen that the payoff for the first entry time into

$$B = (-\infty, 17) \cup (20, \infty)$$

is

$$\mathbb{X}_\tau(\omega) = \begin{cases} 1 & \text{if } \omega \in \{\omega_1, \omega_2, \omega_3, \omega_4\} \\ 0.78 & \text{if } \omega \in \{\omega_5\} \\ 0 & \text{if } \omega \in \{\omega_6, \omega_7, \omega_8\} \end{cases}$$

Hence,

$$\mathcal{E}_\Pi(\mathbb{X}_\tau) = \frac{1}{2} \cdot 1 + \frac{1}{8} \cdot 0.78 + \frac{3}{8} \cdot 0 = 0.5975$$

Consider the stopping time σ defined by

$$\sigma(\omega) = \begin{cases} 2 & \text{if } \omega \in \{\omega_1, \omega_2, \omega_7, \omega_8\} \\ 3 & \text{otherwise} \end{cases}$$

In this case, the (discounted) final value is

$$\mathbb{X}_\sigma(\omega) = \begin{cases} 3.2 & \text{if } \omega \in \{\omega_1, \omega_2\} \\ 0.78 & \text{if } \omega = \{\omega_3, \omega_5\} \\ 0 & \text{if } \omega \in \{\omega_4, \omega_6, \omega_7, \omega_8\} \end{cases}$$

Hence,

$$\mathcal{E}_\Pi(\mathbb{X}_\sigma) = \frac{1}{4} \cdot 3.2 + \frac{1}{4} \cdot 0.78 = 0.995$$

and so σ has a higher fair price than τ and so *in this sense* is a better choice. In fact, as we will see, σ is an optimal stopping time.$\square$

We will compute the Snell envelop of the payoff process from Example 8.1 a bit later, when we have some additional tools that will make the computation easier.

Existence of Optimal Stopping Times

We turn to the question of the existence of optimal stopping times. Let

$$\mathcal{P}_k = \{B_{k,1}, \ldots, B_{k,c}\}$$

be the time-t_k state partition. For $k = 0$, it is clear that optimal stopping times exist because we are simply maximizing a finite set of *constants* $\mathcal{E}_\Pi(\mathbb{X}_\tau)$. But for $k > 0$ we are maximizing nonconstant *functions*

$$\mathcal{E}_\Pi(\mathbb{X}_\tau \mid \mathcal{P}_k) = \sum_{u=1}^c \mathcal{E}_\Pi(\mathbb{X}_\tau \mid B_{k,u}) 1_{B_{k,u}}$$

over stopping times $\tau \in \mathcal{S}_{k,T}$.

However, these are very speical types of functions, namely, linear combinations of indicator functions of a *partition* of Ω, where the coefficients $\mathcal{E}_\Pi(\mathbb{X}_\tau \mid B_{k,u})$ come from a *finite* set, since $\mathcal{S}_{k,T}$ is finite. Such a function is a maximum among all such functions if and only if each coefficient is maximum within its range of values. Thus, for each u, we choose a stopping time $\tau_u \in \mathcal{S}_{k,T}$ for which

$$\mathcal{E}_\Pi(\mathbb{X}_{\tau_u} \mid B_{k,u}) \geq \mathcal{E}_\Pi(\mathbb{X}_\tau \mid B_{k,u}) \tag{8.1}$$

for all $\tau \in \mathcal{S}_{k,T}$. Theorem 8.1 then implies that the linear combination

$$\tau^* = \sum_{u=1}^c \tau_u 1_{B_{k,u}} \in \mathcal{S}_{k,T}$$

is indeed a stopping time and so

$$\mathcal{E}_\Pi(\mathbb{X}_{\tau^*} \mid \mathcal{P}_k) \geq \mathcal{E}_\Pi(\mathbb{X}_\tau \mid \mathcal{P}_k)$$

for all $\tau \in \mathcal{S}_{k,T}$; that is, τ^* is optimal.

Theorem 8.2 *For any interval* $[k, T]$, *an optimal stopping time for* $\mathbb{X}$ *exists. Specifically, if*

$$\mathcal{P}_k = \{B_{k,1}, \ldots, B_{k,c}\}$$

is the time-t_k state partition and if for each u, a stopping time τ_u is chosen so that

$$\mathcal{E}_\Pi(\mathbb{X}_{\tau_u} \mid B_{k,u}) = \max\{\mathcal{E}_\Pi(\mathbb{X}_\tau \mid B_{k,u}) \mid \tau \in \mathcal{S}_{k,T}\}$$

then the stopping time

$$\tau^* = \sum_{u=1}^c \tau_u 1_{B_{k,u}}$$

is optimal; that is,

$$\mathcal{E}_\Pi(\mathbb{X}_{\tau^*} \mid \mathcal{P}_k) = \mathcal{M}_k(\mathbb{X})$$

for all k. $\square$

Since $\mathcal{E}_\Pi(\mathbb{X}_{\tau^*} \mid \mathcal{P}_k)$ is $\mathcal{P}_k$-measurable, the Snell envelope is $\mathbb{F}$-adapted.

Theorem 8.3 *The Snell envelope* $(\mathcal{M}_0(\mathbb{X}), \ldots, \mathcal{M}_T(\mathbb{X}))$ *is $\mathbb{F}$-adapted.* $\square$

Computing the Snell Envelope

Suppose that τ^* is optimal for the interval $[k, T]$ and consider an arbitrary state $B \in \mathcal{P}_k$. Then

$$\mathcal{E}_\Pi(\mathbb{X}_{\tau^*} \mid B) = \max_{\tau \in \mathcal{S}_{k,T}} \{\mathcal{E}_\Pi(\mathbb{X}_\tau \mid B)\} = \frac{1}{\mathbb{P}(B)} \max_{\tau \in \mathcal{S}_{k,T}} \{\mathcal{E}_\Pi(\mathbb{X}_\tau 1_B)\}$$

Thus, an optimal stopping time is one that maximizes the expected value $\mathcal{E}_\Pi(\mathbb{X}_\tau 1_B)$ over $\tau \in \mathcal{S}_{k,T}$.

Now, for each stopping time $\tau \in \mathcal{S}_{k,T}$, there are two possibilities for the stopping event $\{\tau = k\}$: Either $\{\tau = k\}$ contains B or is disjoint from B. Thus, we can write

$$\max_{\tau \in \mathcal{S}_{k,T}} \{\mathcal{E}_\Pi(\mathbb{X}_\tau 1_B)\} = \max\left\{ \max_{\substack{\tau \in \mathcal{S}_{k,T} \\ B \subseteq \{\tau = k\}}} \{\mathcal{E}_\Pi(\mathbb{X}_\tau 1_B)\}, \max_{\substack{\tau \in \mathcal{S}_{k,T} \\ B \cap \{\tau = k\} = \emptyset}} \{\mathcal{E}_\Pi(\mathbb{X}_\tau 1_B)\} \right\}$$

But if $B \subseteq \{\tau = k\}$, then

$$\mathbb{X}_\tau 1_B = X_k 1_B$$

and so

$$\max_{\substack{\tau \in \mathcal{S}_{k,T} \\ B \subseteq \{\tau = k\}}} \{\mathcal{E}_\Pi(\mathbb{X}_\tau 1_B)\} = X_k 1_B$$

On the other hand, if $\{\tau = k\} \cap B = \emptyset$, then if we replace τ by the stopping time τ' that is identical to τ except that it does not stop on $\{\tau = k\}$, in symbols if

$$\tau'(\omega) = \begin{cases} \tau(\omega) & \text{if } \omega \notin \{\tau = k\} \\ T & \text{if } \omega \in \{\tau = k\} \end{cases}$$

then τ' is a stopping time in $\mathcal{S}_{k+1,T}$ and

$$\mathbb{X}_\tau 1_B = \mathbb{X}_{\tau'} 1_B$$

It follows that when computing

$$\max_{\substack{\tau \in \mathcal{S}_{k,T} \\ B \cap \{\tau = k\} = \emptyset}} \{\mathcal{E}_\Pi(\mathbb{X}_\tau 1_B)\}$$

we need consider only stopping times that belong to $\mathcal{S}_{k+1,T}$; that is,

$$\max_{\substack{\tau \in \mathcal{S}_{k,T} \\ B \cap \{\tau = k\} = \emptyset}} \{\mathcal{E}_\Pi(\mathbb{X}_\tau 1_B)\} = \max_{\tau \in \mathcal{S}_{k+1,T}} \{\mathcal{E}_\Pi(\mathbb{X}_\tau 1_B)\}$$

Thus

$$\max_{\tau \in \mathcal{S}_{k,T}} \{\mathcal{E}_\Pi(\mathbb{X}_\tau 1_B)\} = \max\left\{ \mathcal{E}(X_k 1_B),\ \max_{\tau \in \mathcal{S}_{k+1,T}} \{\mathcal{E}_\Pi(\mathbb{X}_\tau 1_B)\} \right\}$$

and dividing by $\mathbb{P}(B)$ gives

$$\mathcal{E}_\Pi(\mathbb{X}_{\tau^*} \mid B) = \max\left\{ X_k(B),\ \max_{\tau \in \mathcal{S}_{k+1,T}} \{\mathcal{E}_\Pi(\mathbb{X}_\tau \mid B)\} \right\}$$

But this holds for each $B \in \mathcal{P}_k$ and so

$$\mathcal{M}_k(\mathbb{X}) = \max\left\{ X_k,\ \max_{\tau \in \mathcal{S}_{k+1,T}} \{\mathcal{E}_\Pi(\mathbb{X}_\tau \mid \mathcal{P}_k)\} \right\}$$

for all $k = 0, \ldots, T$.

We can improve upon this still further with the help of the tower property, which implies that for all $\tau \in \mathcal{S}_{k+1,T}$

$$\mathcal{E}_\Pi(\mathbb{X}_\tau \mid \mathcal{P}_k) = \mathcal{E}_\Pi(\mathcal{E}_\Pi(\mathbb{X}_\tau \mid \mathcal{P}_{k+1}) \mid \mathcal{P}_k) \le \mathcal{E}_\Pi(\mathcal{M}_{k+1}(\mathbb{X}) \mid \mathcal{P}_k)$$

with equality achieved for at least one $\tau \in \mathcal{S}_{k+1,T}$. Hence

$$\max_{\tau \in \mathcal{S}_{k+1,T}} \{\mathcal{E}_\Pi(\mathbb{X}_\tau \mid \mathcal{P}_k)\} = \mathcal{E}_\Pi(\mathcal{M}_{k+1}(\mathbb{X}) \mid \mathcal{P}_k)$$

and so we finally arrive at an important backward recurrence relation for $\mathcal{M}_k(\mathbb{X})$, which will allow us to compute Snell envelopes.

Theorem 8.4 *The Snell envelope satisfies the backward recurrence relation*
1) $\mathcal{M}_T = X_T$
2) *For $0 \leq k < T$,*

$$\mathcal{M}_k(\mathbb{X}) = \max\{X_k, \mathcal{E}_\Pi(\mathcal{M}_{k+1}(\mathbb{X}) \mid \mathcal{P}_k)\} \qquad \square$$

Now we can compute the Snell envelope of the payoff process from Example 8.1.

Example 8.5 Again referring to Example 8.1, let us compute the Snell envelope of the payoff process (X_k), where $X_0 = 0$ and

$$X_3 = \max\{S_3 - 21, 0\} = \begin{cases} 5.62 & \text{for } \omega = \omega_1 \\ 0.78 & \text{for } \omega = \omega_2, \omega_3, \omega_5 \\ 0 & \text{otherwise} \end{cases}$$

$$X_2 = \max\{S_2 - 21, 0\} = \begin{cases} 3.2 & \text{for } \omega \in B_{2,1} \\ 0 & \text{otherwise} \end{cases}$$

$$X_1 = \max\{S_1 - 21, 0\} = \begin{cases} 1 & \text{for } \omega \in B_{1,1} \\ 0 & \text{otherwise} \end{cases}$$

First, we have

$$\mathcal{M}_3(\mathbb{X}) = X_3$$

Next,

$$\mathcal{E}_\Pi(\mathcal{M}_3(\mathbb{X}) \mid \mathcal{P}_2) = \mathcal{E}_\Pi(X_3 \mid \mathcal{P}_2) = \begin{cases} \frac{1}{2}(5.26 + 0.78) = 3.2 & \text{if } \omega \in B_{2,1} \\ \frac{1}{2}(0.78 + 0) = 0.39 & \text{if } \omega \in B_{2,2} \\ \frac{1}{2}(0.78 + 0) = 0.39 & \text{if } \omega \in B_{2,3} \\ 0 & \text{if } \omega \in B_{2,4} \end{cases}$$

and so

$$\mathcal{M}_2(\mathbb{X}) = \max\{X_2, \mathcal{E}_\Pi(\mathcal{M}_3(\mathbb{X}) \mid \mathcal{P}_2)\} = \begin{cases} 3.2 & \text{if } \omega = \omega_1, \omega_2 \\ 0.39 & \text{if } \omega = \omega_3, \omega_4 \\ 0.39 & \text{if } \omega = \omega_5, \omega_6 \\ 0 & \text{if } \omega = \omega_7, \omega_8 \end{cases}$$

Next,

$$\mathcal{E}_\Pi(\mathcal{M}_2(\mathbb{X}) \mid \mathcal{P}_1) = \begin{cases} \frac{1}{2}(3.2 + 0.39) = 1.795 & \text{if } \omega = \omega_1, \omega_2, \omega_3, \omega_4 \\ \frac{1}{2}(0.39 + 0) = 0.195 & \text{if } \omega = \omega_5, \omega_6, \omega_7, \omega_8 \end{cases}$$

and so

$$\mathcal{M}_1(\mathbb{X}) = \max\{X_1, \mathcal{E}_\Pi(\mathcal{M}_2(\mathbb{X}) \mid \mathcal{P}_1)\}$$
$$= \begin{cases} 1.795 & \text{if } \omega = \omega_1, \omega_2, \omega_3, \omega_4 \\ 0.195 & \text{if } \omega = \omega_5, \omega_6, \omega_7, \omega_8 \end{cases}$$

Finally,

$$\mathcal{M}_0(\mathbb{X}) = \max\{X_0, \mathcal{E}_\Pi(\mathcal{M}_1(\mathbb{X}) \mid \mathcal{P}_0)\}$$
$$= \max\left\{0, \frac{1}{2}(1.795 + 0.195)\right\}$$
$$= 0.995$$

Let us recall from Example 8.4 that the stopping time σ defined by

$$\sigma(\omega) = \begin{cases} 2 & \text{if } \omega \in \{\omega_1, \omega_2, \omega_7, \omega_8\} \\ 3 & \text{otherwise} \end{cases}$$

has expected (discounted) final value

$$\mathcal{E}_\Pi(\overline{X}_\sigma) = \frac{1}{4} \cdot 3.2 + \frac{1}{4} \cdot 0.78 = 0.995$$

which is equal to $\mathcal{M}_0(\mathbb{X})$. Hence, σ is indeed an optimal stopping time for the interval $[0, T]$. In fact, as we will see, σ is the *smallest* optimal stopping time in the sense that it stops before any other optimal stopping time. (Observe that we could have waited until time t_3 in states ω_7 and ω_8 and still achieved optimality.)$\square$

This is a good time to emphasize a point about optimal stopping times, namely, optimal stopping times represent the best *guess* as to when to stop without being able to see into the future. Thus, an optimal stopping time is not *guaranteed* to produce the best possible payoff. Indeed, looking at Figure 8.1, it is clear that the best possible exercise procedure involves exercising at time t_2 if the final state is ω_2 but waiting until time t_3 if the final state is ω_1. However, at time t_2 we do not know which state will prevail: ω_1 or ω_2 so this plan is *not* a stopping time.

The Smallest Dominating Supermartingale

It is clear from condition 2) of Theorem 8.4 that

$$\mathcal{E}_\Pi(\mathcal{M}_{k+1}(\mathbb{X}) \mid \mathcal{P}_k) \le \mathcal{M}_k(\mathbb{X})$$

which is the condition that $\mathcal{M}_k(\mathbb{X})$ be a *supermartingale*, as defined below.

Definition *An $\mathbb{F}$-adapted process $(X_0, \ldots, X_T)$ is a* **supermartingale** *if*

$$\mathcal{E}(X_{k+1} \mid \mathcal{P}_k) \le X_k$$

for all k.$\square$

It is also clear that

$$X_k \leq \mathcal{M}_k(\mathbb{X})$$

that is, $\mathcal{M}_k(\mathbb{X})$ **dominates** X_k. It is not hard to see using the recurrence relation that $\mathcal{M}_k(\mathbb{X})$ is the *smallest* supermartingale that dominates X_k.

Theorem 8.5 *The Snell envelope* $(\mathcal{M}_k(\mathbb{X}))$ *is the smallest supermartingale that dominates* (X_k).
Proof. We have seen that $\mathcal{M}_k(\mathbb{X})$ is a supermartingale that dominates X_k. Suppose that W_k is a supermartingale that dominates X_k. This is equivalent to the single inequality

$$W_k \geq \max\{X_k, \mathcal{E}_\Pi(W_{k+1} \mid \mathcal{P}_k)\}$$

We can now proceed by backward induction using the recurrence relation. For the basis step in the induction, we have

$$W_T \geq X_T = \mathcal{M}_T(\mathbb{X})$$

Assuming that $W_{k+1} \geq \mathcal{M}_{k+1}(\mathbb{X})$ then

$$\begin{aligned}
W_k &\geq \max\{X_k, \mathcal{E}_\Pi(W_{k+1} \mid \mathcal{P}_k)\} \\
&\geq \max\{X_k, \mathcal{E}_\Pi(\mathcal{M}_{k+1}(\mathbb{X}) \mid \mathcal{P}_k)\} \\
&= \mathcal{M}_k(\mathbb{X})
\end{aligned}$$

which completes the proof.$\square$

Additional Facts about Martingales

Our next goal is to compute the optimal stopping times that stop as early as possible and the optimal stopping times that stop as late as possible. For this, we need some additional results relating to martingales and supermartingales. Recall (Theorem 4.2) that if $\mathbb{X}$ is a martingale, then for all j and k,

$$\mathcal{E}(X_j) = \mathcal{E}(X_k)$$

We have a corresponding result for supermartingales, whose proof is left as an exercise.

Theorem 8.6
1) If $\mathbb{X}$ *is a martingale then for all j and k,*

$$\mathcal{E}(X_j) = \mathcal{E}(X_k)$$

2) If $\mathbb{X}$ *is a supermartingale, then*

$$j \leq k \quad \Rightarrow \quad \mathcal{E}(X_j) \geq \mathcal{E}(X_k) \qquad\qquad \square$$

Stopping a Process: Doob's Optional Stopping Theorem

A stochastic process $\mathbb{X} = (X_0, \ldots, X_T)$ is defined by its sequences of values

$$X_0(\omega), \ldots, X_T(\omega)$$

for all $\omega \in \Omega$. These sequences have a name.

Definition *If* $\mathbb{X} = (X_0, \ldots, X_T)$ *then for each* $\omega \in \Omega$, *the sequence*

$$X_0(\omega), \ldots, X_T(\omega)$$

is called a **sample path.** $\square$

We can use a stopping time τ to stop the progress of a sample path

$$X_0(\omega), \ldots, X_T(\omega)$$

Specifically, for each $\omega \in \Omega$, we stop the progress of the sample path at time $\tau(\omega)$ and so if $\tau(\omega) = n$, then the stopped sample path is

$$X_0(\omega), \ldots, X_{n-1}(\omega), X_n(\omega), X_n(\omega), \ldots, X_n(\omega)$$

If we set $a \wedge b = \min\{a, b\}$, then this can be written as

$$X_{0 \wedge \tau(\omega)}(\omega), \ldots, X_{(n-1) \wedge \tau(\omega)}(\omega), X_{n \wedge \tau(\omega)}(\omega), \ldots, X_{T \wedge \tau(\omega)}(\omega)$$

and prompts the following definition.

Definition *Let* $\mathbb{X} = (X_0, \ldots, X_T)$ *be an* $\mathbb{F}$-*adapted stochastic process and let* τ *be a stopping time. The* **stopped process** *or* **sampled process** *is defined by*

$$\mathbb{X}_{\wedge \tau} = (X_{0 \wedge \tau}, \ldots, X_{T \wedge \tau})$$ $\square$

Note that $X_{n \wedge \tau}$ can be decomposed as follows:

$$X_{n \wedge \tau} = \sum_{i=0}^{n-1} X_i 1_{\{\tau = i\}} + X_n 1_{\{\tau \geq n\}} \tag{8.2}$$

The next theorem is one of the key results in martingale theory. (A stopping time is also called an *optional* random variable.)

Theorem 8.7 (Doob's Optional Sampling Theorem) *Let* $\mathbb{X}$ *be a martingale (or supermartingale) and* τ *a stopping time. Then the stopped process* $\mathbb{X}_{\wedge \tau}$ *is also a martingale (or supermartingale).*
Proof. For the martingale case, we know that

$$\mathcal{E}(X_n \mid \mathcal{P}_{n-1}) = X_{n-1}$$

and (8.2) gives

$$\mathcal{E}(X_{n\wedge\tau} \mid \mathcal{P}_{n-1}) = \sum_{i=0}^{n-1}\mathcal{E}(X_i 1_{\{\tau=i\}} \mid \mathcal{P}_{n-1}) + \mathcal{E}(X_n 1_{\{\tau\geq n\}} \mid \mathcal{P}_{n-1})$$

But $X_i 1_{\{\tau=i\}}$ and $1_{\{\tau\geq n\}}$ are $\mathcal{P}_{n-1}$-measurable for $i \leq n-1$ and so

$$\mathcal{E}(X_{n\wedge\tau} \mid \mathcal{P}_{n-1}) = \sum_{i=0}^{n-1} X_i 1_{\{\tau=i\}} + 1_{\{\tau\geq n\}}\mathcal{E}(X_n \mid \mathcal{P}_{n-1})$$

$$= \sum_{i=0}^{n-1} X_i 1_{\{\tau=i\}} + 1_{\{\tau\geq n\}} X_{n-1}$$

$$= \sum_{i=0}^{n-2} X_i 1_{\{\tau=i\}} + 1_{\{\tau\geq n-1\}} X_{n-1}$$

$$= X_{(n-1)\wedge\tau}$$

as desired. The proof for a supermartingale is almost identical. $\square$

The Doob Decomposition

Finally, we need the following decomposition result.

Theorem 8.8 (The Doob Decomposition) *Let* $\mathbb{X} = (X_0,\dots,X_T)$ *be an* $\mathbb{F}$-*adapted stochastic process.*
1) *There exist a unique martingale* $\mathbb{M} = (M_0,\dots,M_T)$ *and a unique predictable process* $\mathbb{A} = (A_1,\dots,A_T)$ *for which*

$$X_k = M_k - A_k$$

with $A_0 = 0$.
2) *If* $\mathbb{X}$ *is a supermartingale, then* $\mathbb{A}$ *is nondecreasing; that is,* $A_k \leq A_{k+1}$ *for all* $0 \leq k \leq T-1$.

Proof. For part 1), set $M_0 = X_0$, $A_0 = 0$. For $k \geq 0$, write

$$A_k = M_k - X_k$$

Taking conditional expectations gives

$$\mathcal{E}_\Pi(A_k \mid \mathcal{P}_{k-1}) = \mathcal{E}_\Pi(M_k \mid \mathcal{P}_{k-1}) - \mathcal{E}_\Pi(X_k \mid \mathcal{P}_{k-1})$$

Assume for a moment that $\mathbb{A}$ is predictable and $\mathbb{M}$ is a martingale. Applying the predictability and martingale conditions gives

$$M_k - X_k = A_k = M_{k-1} - \mathcal{E}_\Pi(X_k \mid \mathcal{P}_{k-1})$$

and so

$$M_k - M_{k-1} = X_k - \mathcal{E}_\Pi(X_k \mid \mathcal{P}_{k-1})$$

Since $M_0 = X_0$, we have

$$M_k - X_0 = \sum_{i=1}^{k}(M_i - M_{i-1}) = \sum_{i=1}^{k}[X_i - \mathcal{E}_\Pi(X_i \mid \mathcal{P}_{i-1})]$$

Thus, we are led to taking

$$M_k = \sum_{i=1}^{k}[X_i - \mathcal{E}_\Pi(X_i \mid \mathcal{P}_{i-1})] + X_0$$

and $A_k = M_k - X_k$. We leave it as an exercise to show that these choices do indeed work and that the processes $\mathbb{M}$ and $\mathbb{A}$ *uniquely* satisfy the requirements of the theorem. For part 2), if $\mathbb{X}$ is a supermartingale, then

$$\begin{aligned} M_k - A_k = X_k &\geq \mathcal{E}_\Pi(X_{k+1} \mid \mathcal{P}_k) \\ &= \mathcal{E}_\Pi(M_{k+1} \mid \mathcal{P}_k) - \mathcal{E}_\Pi(A_{k+1} \mid \mathcal{P}_k) \\ &= M_k - A_{k+1} \end{aligned}$$

and so $A_k \leq A_{k+1}$. $\square$

Characterizing Optimal Stopping Times

For simplicity, we will abbreviate $\mathcal{M}_k(\mathbb{X})$ by $\mathcal{M}_k$, since $\mathbb{X}$ will remain fixed throughout the present discussion. If we stop the Snell envelope $\mathbb{S} = (\mathcal{M}_0, \ldots, \mathcal{M}_T)$ using a stopping time $\tau \in \mathcal{S}_{0,T}$, then Doob's optional sampling theorem implies that the stopped process

$$\mathbb{S}_{\wedge \tau} = (\mathcal{M}_{0 \wedge \tau}, \ldots, \mathcal{M}_{T \wedge \tau})$$

is also a supermartingale. Hence, Theorem 8.6 gives

$$j \leq k \quad \Rightarrow \quad \mathcal{E}_\Pi(\mathcal{M}_{k \wedge \tau}) \leq \mathcal{E}_\Pi(\mathcal{M}_{j \wedge \tau})$$

and since $X_k \leq \mathcal{M}_k$ for all k implies that $\mathbb{X}_\tau \leq \mathcal{M}_\tau$, we have

$$\mathcal{E}_\Pi(\mathbb{X}_\tau) \leq \mathcal{E}_\Pi(\mathcal{M}_\tau) = \mathcal{E}_\Pi(\mathcal{M}_{T \wedge \tau}) \leq \cdots \leq \mathcal{E}_\Pi(\mathcal{M}_{0 \wedge \tau}) = \mathcal{M}_0$$

Now, if τ^* is an *optimal* stopping time for $[0, T]$, then

$$\mathcal{M}_0 = \mathcal{E}_\Pi(\mathbb{X}_{\tau^*} \mid \mathcal{P}_0) = \mathcal{E}_\Pi(\mathbb{X}_{\tau^*})$$

and so the previous sequence of inequalities becomes a sequence of equalities (with $\tau = \tau^*$). The first of these equalities

$$\mathcal{E}_\Pi(\mathbb{X}_{\tau^*}) = \mathcal{E}_\Pi(\mathcal{M}_{\tau^*})$$

implies, since $\mathbb{X}_{\tau^*} \leq \mathcal{M}_{\tau^*}$ and Π is strongly positive, that $\mathcal{M}_{\tau^*} = \mathbb{X}_{\tau^*}$ (see the exercises).

Looking further down the chain of equalities, we also see that

$$\mathcal{E}_\Pi(\mathcal{M}_{k \wedge \tau^*}) = \mathcal{E}_\Pi(\mathcal{M}_{(k-1) \wedge \tau^*})$$

and the supermartingale property

$$\mathcal{E}_\Pi(\mathcal{M}_{k \wedge \tau^*} \mid \mathcal{P}_{k-1}) \le \mathcal{M}_{(k-1) \wedge \tau^*}$$

now implies that $(\mathcal{M}_{k \wedge \tau^*})$ is in fact a martingale, since both sides have the same expected value under a strongly positive probability measure.

For the converse, suppose that $\mathbb{X}_\tau = \mathcal{M}_\tau$ and that $\mathcal{M}_{k \wedge \tau}$ is a martingale. Then the sequence of inequalities is a sequence of equalities and in particular,

$$\mathcal{M}_0 = \mathcal{E}_\Pi(\mathbb{X}_\tau)$$

which implies that τ is optimal for the interval $[0, T]$.

We now have a characterization of optimal stopping times.

Theorem 8.9 *A stopping time* $\tau \in \mathcal{S}_{0,T}$ *is optimal if and only if*
1) $\mathcal{M}_\tau = \mathbb{X}_\tau$
2) *The stopped Snell envelope* $\mathbb{S}_{\wedge \tau} = (\mathcal{M}_{k \wedge \tau})$ *is a martingale.* $\square$

Optimal Stopping Times and the Doob Decomposition

We have seen that a stopping time τ is optimal if and only if $\mathbb{X}_\tau = \mathcal{M}_\tau$ and $\mathbb{S}_{\wedge \tau} = (\mathcal{M}_{k \wedge \tau})$ is a martingale. This prompts us to take a closer look at when $\mathbb{S}_{\wedge \tau}$ is a martingale.

We have seen that the Snell envelope

$$\mathbb{S} = (\mathcal{M}_0, \dots, \mathcal{M}_T)$$

is a supermartingale. Using Doob's decomposition, we can write

$$\mathbb{S} = \mathbb{M} - \mathbb{A}$$

where $\mathbb{M} = (M_0, \dots, M_T)$ is a martingale, $\mathbb{A} = (A_1, \dots, A_T)$ is predictable and nondecreasing and $A_0 = 0$. If we stop this sequence,

$$\mathbb{S}_{\wedge \tau} = \mathbb{M}_{\wedge \tau} - \mathbb{A}_{\wedge \tau}$$

then $\mathbb{M}_{\wedge \tau}$ is a martingale and so it follows from the uniqueness of the Doob decomposition that $\mathbb{S}_{\wedge \tau}$ is a martingale if and only if $\mathbb{A}_{\wedge \tau}$ is the zero process.

Since $\mathbb{A}$ is nondecreasing, so is $\mathbb{A}_{\wedge \tau}$ and so it is the zero process if and only if the final term $A_{T \wedge \tau} = A_\tau$ is the zero random variable. Thus, $\mathbb{S}_{\wedge \tau}$ is a martingale if and only if $A_\tau = 0$.

Theorem 8.10 *Let* $\mathbb{S} = (\mathcal{M}_k)$ *be the Snell envelope of* (X_k). *For a stopping time* $\tau \in \mathcal{S}_{0,T}$, *the stopped process* $\mathbb{S}_{\wedge \tau} = (\mathcal{M}_{k \wedge \tau})$ *is a martingale if and only if* $A_\tau = 0$, *where* $\mathbb{A} = (A_k)$ *is the predictable process in the Doob decomposition of* $\mathbb{S}$. $\square$

The Smallest Optimal Stopping Time

The previous theorem makes it easy to determine the smallest optimal stopping time. First, we recall the recurrence formula

$$\mathcal{M}_k = \max\{X_k, \mathcal{E}_\Pi(\mathcal{M}_{k+1} \mid \mathcal{P}_k)\}$$

Using the Doob decomposition, we notice that

$$\begin{aligned}
\mathcal{E}_\Pi(\mathcal{M}_{k+1} \mid \mathcal{P}_k) &= \mathcal{E}_\Pi(M_{k+1} \mid \mathcal{P}_k) - \mathcal{E}_\Pi(A_{k+1} \mid \mathcal{P}_k) \\
&= M_k - A_{k+1} \\
&= (M_k - A_k) - (A_{k+1} - A_k) \\
&= \mathcal{M}_k - (A_{k+1} - A_k)
\end{aligned}$$

and so

$$\mathcal{M}_k = \max\{X_k, \mathcal{M}_k - (A_{k+1} - A_k)\}$$

Hence, the *strict* inequality

$$\mathcal{M}_k > X_k$$

implies that

$$A_{k+1} = A_k$$

Hence, if $\mathcal{M}_k = X_k$ but $\mathcal{M}_i > X_i$ for all $i < k$, then $A_{i+1} = A_i$ for $i < k$. But $A_0 = 0$ and so $A_i = 0$ for all $i \le k$. This prompts us to define $\tau_{\min}$ by

$$\tau_{\min}(\omega) = \min\{k \mid X_k(\omega) = \mathcal{M}_k(\omega)\}$$

which exists since $X_T(\omega) = \mathcal{M}_T(\omega)$. In addition, $\tau_{\min}$ is a stopping time since it is the first entry time of the adapted process $(X_k - \mathcal{M}_k)$ into the Borel set $\{0\}$. By definition,

$$X_{\tau_{\min}} = \mathcal{M}_{\tau_{\min}}$$

We have just seen that $A_{\tau_{\min}(\omega)}(\omega) = 0$ for all $\omega \in \Omega$ and so $A_{\tau_{\min}} = 0$, which implies that $\mathbb{V}^{\tau_{\min}}$ is a martingale. Hence, Theorem 8.9 implies that $\tau_{\min}$ is an optimal stopping time. Moreover, $\tau_{\min}$ is the *smallest* optimal stopping time because all optimal stopping times τ satisfy $X_{\tau(\omega)}(\omega) = \mathcal{M}_{\tau(\omega)}(\omega)$ for all $\omega \in \Omega$ and so $\tau_{\min}(\omega) \le \tau(\omega)$ for all $\omega \in \Omega$.

Theorem 8.11 *The smallest optimal stopping time is*

$$\tau_{\min}(\omega) = \min\{k \mid X_k(\omega) = \mathcal{M}_k(\omega)\} \qquad \square$$

Example 8.6 In Example 8.4 we defined the stopping time

$$\sigma(\omega) = \begin{cases} 2 & \text{if } \omega \in \{\omega_1, \omega_2, \omega_7, \omega_8\} \\ 3 & \text{otherwise} \end{cases}$$

and showed in Example 8.5 that σ is optimal. In fact, it is easy to see that σ has the property

$$\sigma(\omega) = \min\{k \mid X_k(\omega) = \mathcal{M}_k(\omega)\}$$

and so it is the smallest optimal stopping time.□

The Largest Optimal Stopping Time

In view of Theorem 8.10, in casting about for the largest optimal stopping time, it is natural to consider the function

$$\tau_{\max}(\omega) = \max\{k \mid A_k(\omega) = 0\}$$

which exists since $A_0 = 0$. Since any optimal stopping time τ satisfies $A_\tau = 0$, if $\tau_{\max}$ is an optimal stopping time then it must be the largest optimal stopping time. Note that $\tau_{\max}$ can also be defined by

$$\tau_{\max}(\omega) = \begin{cases} \min\{k \mid A_{k+1}(\omega) > 0\} & \text{if } \{k \mid A_{k+1}(\omega) > 0\} \neq \emptyset \\ T & \text{otherwise} \end{cases}$$

and since this is the first entry time into $(0, \infty)$ of the adapted process $\mathbb{A}$, it follows that $\tau_{\max}$ is a stopping time. Also, since $A_{\tau_{\max}} = 0$, Theorem 8.10 implies that $(\mathcal{M}_k^{\tau_{\max}})$ is a martingale. Thus, to show that $\tau_{\max}$ is optimal, we need only show that $\mathcal{M}_{\tau_{\max}} = X_{\tau_{\max}}$.

Once again we look at the recurrence relation

$$\mathcal{M}_k = \max\{X_k, \mathcal{E}_\Pi(\mathcal{M}_{k+1} \mid \mathcal{P}_k)\}$$

which, using Doob's decomposition can be written

$$\mathcal{M}_k = \max\{X_k, \mathcal{M}_k - (A_{k+1} - A_k)\}$$

But for $k = \tau_{\max}(\omega)$ we have

$$A_{\tau_{\max}+1}(\omega) - A_{\tau_{\max}}(\omega) = A_{\tau_{\max}+1}(\omega) > 0$$

and so the maximum above is just X_k; that is,

$$\mathcal{M}_{\tau_{\max}}(\omega) = X_{\tau_{\max}}(\omega)$$

Thus $\mathcal{M}_{\tau_{\max}} = X_{\tau_{\max}}$ as desired.

Theorem 8.12 *The largest optimal stopping time is*

$$\tau_{\max}(\omega) = \max\{k \mid A_k(\omega) = 0\}$$
$$= \begin{cases} \min\{k \mid A_{k+1}(\omega) > 0\} & \text{if } \{k \mid A_{k+1}(\omega) > 0\} \neq \emptyset \\ T & \text{otherwise} \end{cases}$$

where

$$\mathcal{M}_k = M_k - A_k$$

is the Doob decomposition. □

Exercises

1. Show that the first entry time into any set of the form (a, b) is a stopping time.
2. Show that the maximum of stopping times in $\mathcal{S}_{k,T}$ is a stopping time in $\mathcal{S}_{k,T}$.
3. Show that the first entry time into any Borel set B is a stopping time.
4. Show that the first time that a stock's price doubles its initial price is a stopping time.
5. Show that the first time that a stock's price doubles its previous price is a stopping time; that is, the random variable

    $$\tau(\omega) =$$
    $$\begin{cases} \min\{k \mid S_k(\omega) \geq 2S_{k-1}(\omega)\} & \text{if } \{k \mid S_k(\omega) \geq 2S_{k-1}(\omega)\} \neq \emptyset \\ T & \text{otherwise} \end{cases}$$

    is a stopping time.
6. Show that the first *exit* time from a set of the form (a, b) is a stopping time.
7. Prove that the maximum, minimum or sum of two stopping times is a stopping time. How about the difference?
8. Prove that for any random variables X and Y

    $$\max\{\mathcal{E}(X \mid \mathcal{P}), \mathcal{E}(X \mid \mathcal{P})\} \leq \mathcal{E}(\max\{X, Y\} \mid \mathcal{P})$$

9. Prove that since $X_k \leq \mathcal{M}_k$ for all k, it follows that $\mathbb{X}_\tau \leq \mathcal{M}_\tau$ for any stopping time τ.
10. Prove that if two discrete random variables satisfy $X \leq Y$ and $\mathcal{E}_\Pi(X) = \mathcal{E}_\Pi(Y)$ then because Π is strongly positive, it follows that $X = Y$.
11. Prove that if $(A_0, \dots, A_T)$ is a martingale and $(A_1, \dots, A_T)$ is predictable then (A_k) is a constant sequence; that is,

    $$A_0^\tau = \cdots = A_T^\tau$$

12. Let $\mathbb{X} = (X_0, \dots, X_T)$ be an $\mathbb{F}$-adapted stochastic process. Finish the proof of Doob's decomposition theorem; that is, show that the processes $M_0 = X_0$, $A_0 = 0$ and for $k > 0$,

$$M_k = \sum_{i=1}^{k} [X_i - \mathcal{E}_\Pi(X_i \mid \mathcal{P}_{i-1})] + X_0$$

and $A_k = M_k - X_k$ satisfy the conditions of the decomposition.

13. For the binomial model with

$$u = 1.2, d = 0.8, r = 0, S_0 = 20$$

compute the price process, payoff process for an American call with $K = 21$ and the Snell envelope. Find the first optimal stopping time.

14. Write an Excel spreadsheet that given u, d, r, S_0 and K will compute the price process, payoff process for an American call/put with strike price K and the Snell envelope.

Part 3—The Black–Scholes Option Pricing Formula

Chapter 9
Continuous Probability

In this chapter, we discuss some concepts of the general theory of probability, without restriction on the size of the sample space. This is in preparation for our discussion of the Black–Scholes derivative pricing model.

Since this is not a book on probability and since a detailed discussion of probability would take us too far from our main goals, we will need to be a bit sketchy at times, omitting proofs that would take us too far from our intended scope. For a more complete treatment of probability, please consult the references at the end of this book.

General Probability Spaces

In defining arbitrary probability spaces $(\Omega, \mathbb{P})$ where Ω need not be a finite set, we must require that the *countable* union of a sequence $A_1, A_2, \ldots$ of events,

$$\bigcup_{i=1}^{\infty} A_i$$

also be an event. Moreover, finite additivity must be replaced by *countable additivity*; that is, if $A_1, A_2, \ldots$ is a sequence of pairwise disjoint events in Ω, then

$$\mathbb{P}\left(\bigcup_{i=1}^{\infty} A_i\right) = \sum_{i=1}^{\infty} \mathbb{P}(A_i)$$

Accordingly, we have the following definitions.

Definition *Let Ω be a nonempty set. A nonempty collection Σ of subsets of Ω is a σ-algebra if*
1) $\Omega \in \Sigma$

2) Σ *is closed under* countable *unions; that is, if* $A_1, A_2, \ldots$ *is a sequence of elements of* Σ *then*

$$\bigcup_{i=1}^{\infty} A_i \in \Sigma$$

3) Σ *is closed under complements; that is, if* $A \in \Sigma$ *then* $A^c \in \Sigma$. $\square$

Note that $\emptyset = \Omega^c \in \Sigma$. Also, DeMorgan's laws show that Σ is closed under countable intersections. Now we can define general probability spaces.

Definition *A* **probability space** *is a triple* $(\Omega, \Sigma, \mathbb{P})$ *consisting of a nonempty set* Ω, *called a* **sample space**, *a* σ-*algebra* Σ *of subsets of* Ω *whose elements are called* **events** *and a real-valued function* $\mathbb{P}$ *defined on* Σ *called a* **probability measure**. *The function* $\mathbb{P}$ *must satisfy the following properties:*
1)   *(Range) For all* $A \in \Sigma$,

$$0 \leq \mathbb{P}(A) \leq 1$$

2)   *(Probability of* Ω)

$$\mathbb{P}(\Omega) = 1$$

3)   *(Countable additivity) If*

$$A_1, A_2, \ldots$$

is a sequence of pairwise mutually exclusive events, then

$$\mathbb{P}\left(\bigcup_{i=1}^{\infty} A_i\right) = \sum_{i=1}^{\infty} \mathbb{P}(A_i) \qquad \qquad \square$$

Continuity of Probability Measures

A **decreasing sequence** of events is a sequence of events satisfying

$$A_1 \supseteq A_2 \supseteq \cdots$$

Similarly, an **increasing sequence** of events is a sequence of events satisfying

$$A_1 \subseteq A_2 \subseteq \cdots$$

Theorem 9.1 *Probability measures are* **monotonically continuous** *in the following sense:*
1)   *If* $A_1 \supseteq A_2 \supseteq \cdots$ *is a decreasing sequence of events, then*

$$\lim_{i \to \infty} \mathbb{P}(A_i) = \mathbb{P}\left(\bigcap_{i=1}^{\infty} A_i\right)$$

2) *If* $A_1 \subseteq A_2 \subseteq \cdots$ *is an* increasing *sequence of events, then*

$$\lim_{i \to \infty} \mathbb{P}(A_i) = \mathbb{P}\left(\bigcup_{i=1}^{\infty} A_i\right)$$

Proof. For part 1), the sequence $\mathbb{P}(A_i)$ of probabilities is nonincreasing and is bounded below by 0. It is a theorem of elementary real analysis that such a sequence must converge, so the limit $\lim \mathbb{P}(A_i)$ does exist. For convenience, let $A = \bigcap_{i=1}^{\infty} A_i$. As shown in Figure 9.1,

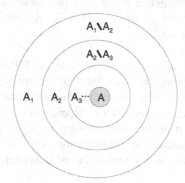

Figure 9.1

the events

$$A_1 \setminus A_2, A_2 \setminus A_3, \ldots$$

are pairwise disjoint and also disjoint from the intersection A. Thus, A_1 is the *disjoint* union

$$A_1 = \left(\bigcup_{i=1}^{\infty}(A_i \setminus A_{i+1})\right) \cup A$$

and so

$$
\begin{aligned}
\mathbb{P}(A_1) &= \sum_{i=1}^{\infty} \mathbb{P}(A_i \setminus A_{i-1}) + \mathbb{P}(A) \\
&= \lim_{n \to \infty} \sum_{i=1}^{n} \mathbb{P}(A_i \setminus A_{i-1}) + \mathbb{P}(A) \\
&= \lim_{n \to \infty} \mathbb{P}(A_1 \setminus A_{n-1}) + \mathbb{P}(A) \\
&= \lim_{n \to \infty} (\mathbb{P}(A_1) - \mathbb{P}(A_{n-1})) + \mathbb{P}(A) \\
&= \mathbb{P}(A_1) - \lim_{n \to \infty} \mathbb{P}(A_{n-1}) + \mathbb{P}(A)
\end{aligned}
$$

and so

$$\lim_{n\to\infty} \mathbb{P}(A_n) = \lim_{n\to\infty} \mathbb{P}(A_{n-1}) = \mathbb{P}(A)$$

as desired. We leave proof of part 2) as an exercise. $\square$

Probability Measures on $\mathbb{R}$

The only nonfinite probability space that we will need to consider in this book is the set $\Omega = \mathbb{R}$ of real numbers. The most important σ-algebra on $\mathbb{R}$ is the *Borel σ-algebra*, defined as follows.

Definition *The* **Borel σ-algebra** $\mathcal{B}$ *on* $\mathbb{R}$ *is the smallest σ-algebra on $\mathbb{R}$ that contains all open intervals* (a,b) *for* $a,b \in \mathbb{R}$. *An element of the Borel σ-algebra is called a* **Borel set**. $\square$

To see that the Borel σ-algebra exists, we note the following. First, there a σ-algebra on $\mathbb{R}$ that contains the open intervals, namely, the collection of all subsets of $\mathbb{R}$. Second, it is easy to prove that the intersection of σ-algebras is also a σ-algebra. Hence, the intersection of all σ-algebras that contain the open intervals is the smallest σ-algebra that contains the open intervals and so is the Borel σ-algebra. Unfortunately, this description is not very practical. However, the following theorem describes some important Borel sets and is all that we will need.

Theorem 9.2
1) *All open, closed and half-open intervals are Borel sets.*
2) *All rays* $(-\infty, b]$, $(-\infty, b)$, $[a, \infty)$ *and* (a, ∞) *are Borel sets.*
3) *All open sets and all closed sets are Borel sets.*
Proof. We sketch the proof. For 1), to see that the half-open interval $(a, b]$ is a Borel set observe that

$$(a,b] = \bigcap_{n=1}^{\infty} \left(a, b + \frac{1}{n} \right)$$

and so $(a, b]$ is the countable union of open intervals and is therefore in $\mathcal{B}$. We leave proof of part 2) to the reader.

For part 3), we need to make a few comments about open sets in $\mathbb{R}$. First, a subset A of $\mathbb{R}$ is **open** if for every $x \in A$ there is an open interval (a, b) for which

$$x \in (a, b) \subseteq A$$

A set is **closed** if its complement is open.

An open interval $I = (a, b)$ in A is **maximal** in A if no open interval containing I as a proper subset is also in A. Any two distinct maximal open intervals are disjoint, since otherwise their union would be a strictly larger open interval contained in A. Also, every maximal open interval contains a *distinct* rational number and since there are only a countable number of rational numbers, there are at most a countable number of maximal open intervals.

Now, every open set A in $\mathbb{R}$ is the union of all maximal open intervals contained within A. Hence, A is the union of at most a countable number of open intervals and is therefore a Borel set. Moreover, since all open sets are Borel sets and since a closed set is the complement of an open set, all closed sets are also Borel sets.□

At first, the more one thinks about Borel sets, the more one comes to feel that all subsets of $\mathbb{R}$ are Borel sets. However, this is not the case, but it is very difficult to describe a set that is not a Borel set. Unfortunately, this is a matter that lies outside the scope of this book.

From now on, the phrase "let $\mathbb{P}$ be a probability measure on $\mathbb{R}$" will carry with it the tacit understanding that the σ-algebra involved is the Borel σ-algebra. We omit proof of the following.

Theorem 9.3 *A probability measure on $\mathbb{R}$ is uniquely determined by its values on the rays $(-\infty, t]$. That is, if $\mathbb{P}$ and $\mathbb{Q}$ are probability measures on $\mathbb{R}$ and*

$$\mathbb{P}((-\infty, t]) = \mathbb{Q}((-\infty, t])$$

for all $t \in \mathbb{R}$ then $\mathbb{P} = \mathbb{Q}$.□

Distribution Functions

For finite or discrete probability spaces, probability measures are most easily described via their mass functions

$$f(\omega) = \mathbb{P}(\{\omega\})$$

However, the concept of a mass function is not general enough to describe all probability measures on arbitrary sample spaces. For this, we need the following concept.

Definition A **distribution function** *is a function $F \colon \mathbb{R} \to \mathbb{R}$ with the following properties:*
1) *F is **nondecreasing**; that is,*

$$s < t \quad \Rightarrow \quad F(s) \leq F(t)$$

   (Note that some authors use the term increasing *for this property.)*
2) *F is **right-continuous**; that is, the right-hand limit exists everywhere and*

$$\lim_{t \to a+} F(t) = F(a)$$

3)   F satisfies

$$\lim_{t \to -\infty} F(t) = 0 \quad and \quad \lim_{t \to \infty} F(t) = 1 \qquad \square$$

Figure 9.2 shows the graph of a probability distribution function. Note that the function in this example is right-continuous but not continuous.

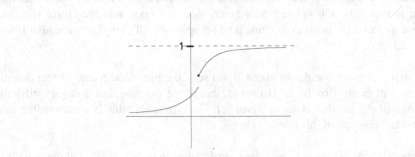

Figure 9.2: A probability distribution function

The extreme importance of probability distribution functions is described in the next theorem. Basically, it implies that there is a one-to-one correspondence between probability measures on $\mathbb{R}$ and probability distribution functions. Thus, knowing either one uniquely determines the other and so the two concepts are essentially equivalent.

Theorem 9.4
1)   Let $\mathbb{P}$ be a probability measure on $\mathbb{R}$. The function $F_\mathbb{P}: \mathbb{R} \to \mathbb{R}$ defined by

$$F_\mathbb{P}(t) = \mathbb{P}((-\infty, t])$$

is a probability distribution function, called the **distribution function** *of $\mathbb{P}$.*
2)   Let $F: \mathbb{R} \to \mathbb{R}$ be a distribution function. Then there is a unique probability measure $\mathbb{P}_F$ on $\mathbb{R}$ whose distribution function is F; that is, for which

$$\mathbb{P}_F((-\infty, t]) = F(t) \qquad \square$$

If $\mathbb{P}$ is a probability measure with distribution function $F_\mathbb{P}$, then we can form the probability measure $\mathbb{Q}$ of $F_\mathbb{P}$. According to the definitions,

$$\mathbb{Q}((-\infty, t]) = F_\mathbb{P}(t) = \mathbb{P}((-\infty, t])$$

and so Theorem 9.3 implies that $\mathbb{P} = \mathbb{Q}$. Also, if F is a distribution function with probability measure $\mathbb{P}_F$, then the distribution function G of $\mathbb{P}_F$ satisfies

$$G(t) = \mathbb{P}_F((-\infty, t]) = F(t)$$

Thus, the correspondences

$$\mathbb{P} \to F_\mathbb{P} \quad and \quad F \to \mathbb{P}_F$$

are inverses of one another and so the notions of probability measure and distribution function are just two different ways of looking at the same concept.

Example 9.1 To convey the notion of equal likelihood for a finite sample space $\{1, \ldots, n\}$, we simply assign the same probabilities $1/n$ to each elementary event $\{k\}$. However, to convey equal likelihood for an infinite sample space such as the closed interval $[0, 1]$ in $\mathbb{R}$, we cannot simply assign a positive real number p to each elementary event $\{r\}$ for all $r \in [0, 1]$ because there are an infinite number of elementary events and so the sum of their probabilities is not finite, let alone equal to 1. Indeed, the probability of each elementary event must be 0 and we must turn our attention to more complicated Borel sets.

To convey the notion of equal likelihood, it seems reasonable that two subintervals of $[0, 1]$ of equal length should have equal probability. It follows, for example, that the interval $[0, 1/2]$ must have probability $1/2$ and more generally that the intervals $[0, 1/n]$ must have probability $1/n$. In fact, the intervals $[0, m/n]$ must have probability m/n for all $0 \leq m/n \leq 1$. Hence, the distribution function F for $\mathbb{P}$ satisfies

$$F(t) = \begin{cases} 0 & \text{if } t < 0 \\ t & \text{if } 0 \leq t \leq 1 \\ 1 & \text{if } t > 1 \end{cases}$$

at least for all rational $t = m/n$. But F must be right continuous and so this must hold for all real numbers t. This distribution function is called the **uniform distribution function on** $[0, 1]$. Figure 9.3 shows the graph of F.□

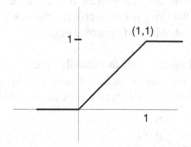

Figure 9.3: The uniform distribution function for $[0, 1]$

Example 9.2 The most important of all probability distributions is the **normal distribution**, whose distribution function is

$$\phi_{\mu,\sigma}(t) = \frac{1}{\sqrt{2\pi\sigma^2}} \int_{-\infty}^{t} e^{-\frac{(x-\mu)^2}{2\sigma^2}} \, dx$$

Figure 9.4 shows the normal distribution function.

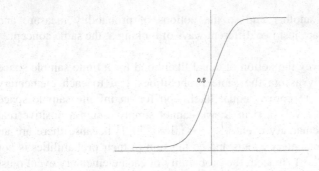

Figure 9.4: The normal distribution function

It can be shown that the parameters μ and σ^2 are the mean and variance of the distribution, respectively. The **standard normal distribution** is the normal distribution with mean 0 and variance 1 and thus has distribution function

$$\phi_{0,1}(t) = \frac{1}{\sqrt{2\pi}} \int_{-\infty}^{t} e^{-\frac{x^2}{2}} dx \qquad \square$$

Note that the uniform and the normal distribution functions are both continuous, not just right-continuous. Put another way, their graphs have no jumps. A jump in the distribution function indicates a point at which the probability is not 0. Let us illustrate with an example.

Example 9.3 A public drug manufacturing company has a new drug that is awaiting FDA approval. If the drug is approved, the company estimates that its stock will end trading that day somewhere in the range $[10, 15]$, with uniform likelihood. However, if the drug is not approved, the stock price will likely be 5. Let us assume that the probability of approval is 0.75.

We could model this situation with the sample space $\Omega = \{5\} \cup [10, 15]$ but it may be simpler to use the sample space $\mathbb{R}$ and simply assign a 0 probability outside the set Ω. The distribution function for this probability measure is

$$F(t) = \begin{cases} 0 & t < 5 \\ 0.25 & 5 \leq t < 10 \\ 0.25 + 0.75\left(\frac{t-10}{5}\right) & 10 \leq t < 15 \\ 1 & t \geq 15 \end{cases}$$

The graph is shown in Figure 9.5. Note the jump at $t = 5$. $\square$

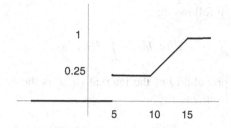

Figure 9.5: A distribution function with a jump

Density Functions

The distribution function of a probability measure is not always the simplest way to describe a probability measure. Many probability measures that occur in applications have the property that their distribution functions are differentiable and that their derivatives are very well-behaved.

By well-behaved, we mean that the derivative of the distribution function F is integrable and the integral is again equal to F. (There are functions that have derivatives that are integrable, but the integral of the derivative is not the original function.) This can be expressed in symbols as follows:

$$F(t) = \int_{-\infty}^{t} F'(x)\, dx$$

The function $f(x) = F'(x)$ is called a *density function* for $\mathbb{P}$. Let us have a formal definition.

Definition

*1) A **density function** $f: \mathbb{R} \to \mathbb{R}$ is a nonnegative function for which*

$$\int_{-\infty}^{\infty} f(x)\, dx = 1$$

that is, the area under the entire graph of f over the entire x-axis is equal to 1.

*2) A probability measure $\mathbb{P}$ on $\mathbb{R}$ or equivalently a distribution function $F_{\mathbb{P}}$ is **absolutely continuous** if there is a density function $f: \mathbb{R} \to \mathbb{R}$ for which*

$$F_{\mathbb{P}}(t) = \mathbb{P}((-\infty, t]) = \int_{-\infty}^{t} f(x)\, dx$$

(Recall that the half-rays are sufficient to define the probability measure.)□

From this definition, it follows that

$$\mathbb{P}((a,b]) = \int_a^b f(x)\, dx$$

In other words, the probability of the interval $(a,b]$ is the *area under graph of the density function* from a to b.

Absolutely continuous probability measures are rather special. For example, their distribution functions are continuous (not just right-continuous). Thus, all elementary events have probability 0.

Example 9.4 The uniform distribution function on $[0,1]$ is absolutely continuous, with density function

$$f(x) = \begin{cases} 0 & x \notin [0,1] \\ 1 & x \in [0,1] \end{cases}$$

The graph of f is shown in Figure 9.6.$\square$

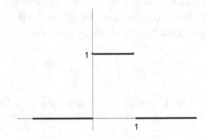

Figure 9.6: The uniform density function on $[0,1]$

Example 9.5 The normal distribution is absolutely continuous, with density function

$$f(x) = \frac{1}{\sqrt{2\pi\sigma^2}} e^{-\frac{(x-\mu)^2}{2\sigma^2}}$$

The density of the standard normal distribution is

$$f(x) = \frac{1}{\sqrt{2\pi}} e^{-\frac{x^2}{2}}$$

This is pictured in Figure 9.7. The graph of the normal density function is the oft spoken of *bell-shaped curve.*$\square$

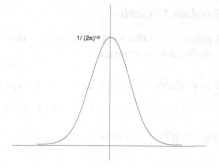

$1/(2\pi)^{1/2}$

Figure 9.7: The standard normal density function

Random Variables

Just as the issue of events is more complex in the nonfinite case, so is the notion of random variable. In particular, not all functions are random variables.

Definition *Let Σ be a σ-algebra on a nonempty set Ω. A function $X: \Omega \to \mathbb{R}$ is Σ-**measurable** if the inverse image of every open interval is in Σ, in symbols*

$$X^{-1}((a,b)) \in \Sigma$$

*A measurable function on (Ω, Σ) is also called a **random variable**.* $\square$

Here are a few facts about random variables, whose proofs we omit.

Theorem 9.5
1) *The sum and product of random variables are random variables, as is any constant multiple of a random variable.*
2) *The composition of random variables is a random variable.*
3) *Continuous and piecewise continuous functions are random variables.* $\square$

The Distribution Function of a Random Variable

If $(\Omega, \Sigma, \mathbb{P})$ is a probability space and X is a random variable on (Ω, Σ), then X defines a distribution function F_X and corresponding probability measure $\mathbb{P}_X$ on $\mathbb{R}$ by

$$F_X(t) = \mathbb{P}_X((-\infty, t]) = \mathbb{P}(X \le t)$$

If $\mathbb{P}_X$ is finite, discrete or absolutely continuous, then we say that the *random variable* is **finite**, **discrete** or **absolutely continuous**, respectively. Absolutely continuous random variables are often simply called **continuous random variables**.

Independence of Random Variables

The definition of independence is the same for arbitrary random variables as it is for random variables on finite sample spaces.

Definition *Two random variables X and Y on $\mathbb{R}$ are* **independent** *if*

$$\mathbb{P}(X \le t, Y \le s) = \mathbb{P}(X \le t)\mathbb{P}(Y \le s)$$

for all $s, t \in \mathbb{R}$. More generally, a collection $X_1, \ldots, X_n$ of random variables is **independent** *if*

$$\mathbb{P}(X_1 \le t_1, \ldots, X_n \le t_n) = \prod_{i=1}^{n} \mathbb{P}(X_i \le t_i) \qquad \square$$

This definition expresses formally the feeling that if random variables are independent then knowledge of the value of one random variable does not affect the probabilities associated with the other random variable.

Expectation and Variance of a Random Variable

Recall that for a random variable X on a finite probability space $(\Omega, \mathbb{P})$ with $\Omega = \{\omega_1, \ldots, \omega_n\}$, the expected value (or mean) is defined by

$$\mathcal{E}_{\mathbb{P}}(X) = \sum_{i=1}^{n} X(\omega_i)\mathbb{P}(\omega_i)$$

If $g \colon \mathbb{R} \to \mathbb{R}$ is a function then the expected value of the random variable $g(X)$ is

$$\mathcal{E}_{\mathbb{P}}(g(X)) = \sum_{i=1}^{n} g(X(\omega_i))\mathbb{P}(\omega_i)$$

Also, the variance is defined by

$$\mathrm{Var}(X) = \mathcal{E}((X - \mu)^2)$$

Let us now extend these concepts to absolutely continuous random variables.

Definition *Let X be an absolutely continuous random variable, with density function f. The* **expected value** *or* **mean** *of X is the improper integral*

$$\mathcal{E}(X) = \int_{-\infty}^{\infty} x f(x)\, dx$$

which exists provided that

$$\int_{-\infty}^{\infty} |x| f(x)\, dx < \infty$$

The **variance** *of X is*

$$\text{Var}(X) = \mathcal{E}((X - \mu)^2)$$

and the **standard deviation** *is the positive square root of the variance*

$$\sigma_X = \sqrt{\text{Var}(X)} \qquad \square$$

Also, if $g: \mathbb{R} \to \mathbb{R}$ is a measurable function, then the random variable $g(X)$ has expected value

$$\mathcal{E}(g(X)) = \int_{-\infty}^{\infty} g(x)f(x)\,dx$$

provided that

$$\int_{-\infty}^{\infty} |g(x)|f(x)\,dx < \infty$$

Here are some basic properties of expectation and variance.

Theorem 9.6

1)   The expected value operator is linear; that is,

$$\mathcal{E}(aX + bY) = a\mathcal{E}(X) + b\mathcal{E}(Y)$$

2)   If $X_1, \ldots, X_n$ are independent random variables on $\mathbb{R}$, then

$$\mathcal{E}(X_1 \cdots X_n) = \prod_{i=1}^{n} \mathcal{E}(X_i)$$

3) $\text{Var}(X) = \mathcal{E}(X^2) - \mu^2 = \mathcal{E}(X^2) - \mathcal{E}(X)^2$
4)   For any real number a

$$\text{Var}(aX) = a^2\text{Var}(X)$$

and

$$\text{Var}(X - a) = \text{Var}(X)$$

5)   If $X_1, \ldots, X_n$ are independent random variables on $\mathbb{R}$, then

$$\text{Var}(X_1 + \cdots + X_n) = \sum_{i=1}^{n} \text{Var}(X_i) \qquad \square$$

The Normal Distribution

Let us take another look at the normal distribution, whose density function is

$$N_{\mu,\sigma}(x) = \frac{1}{\sqrt{2\pi\sigma^2}} e^{-\frac{(x-\mu)^2}{2\sigma^2}}$$

We mentioned that the parameters μ and σ^2 are the mean and variance, respectively. To calculate the mean, we need only a bit of first-year calculus. The definition is

$$\mathcal{E} = \frac{1}{\sqrt{2\pi\sigma^2}} \int_{-\infty}^{\infty} x e^{-\frac{(x-\mu)^2}{2\sigma^2}} dx$$

Writing $x = (x - \mu) + \mu$ and splitting the integral gives

$$\mathcal{E} = \frac{1}{\sqrt{2\pi\sigma^2}} \int_{-\infty}^{\infty} (x-\mu) e^{-\frac{(x-\mu)^2}{2\sigma^2}} dx + \frac{\mu}{\sqrt{2\pi\sigma^2}} \int_{-\infty}^{\infty} e^{-\frac{(x-\mu)^2}{2\sigma^2}} dx$$

The second integral is just μ times the integral of $N_{\mu,\sigma}$ and since this integral is 1, we just get μ. As to the first integral, the substitution $y = x - \mu$ gives

$$\int_{-\infty}^{\infty} y e^{-\frac{y^2}{2\sigma^2}} dy$$

But the integrand $y e^{-y^2/2\sigma^2}$ is an odd function, from which it follows that the integral from $-\infty$ to ∞ must be 0. (We leave elaboration of this as an exercise.) Hence, $\mathcal{E} = 0 + \mu = \mu$.

Computation of the variance of the normal distribution requires the beautiful but nontrivial integral formula

$$\int_{-\infty}^{\infty} y^2 e^{-\frac{y^2}{2}} dy = \sqrt{2\pi}$$

From here, the rest is straightforward, especially using the formula

$$\text{Var}(X) = \mathcal{E}(X^2) - \mathcal{E}(X)^2$$

The upshot is that $\text{Var}(N_{\mu,\sigma}) = \sigma^2$. We leave the details as an exercise.

If $N_{\mu,\sigma}$ is a normal random variable with mean μ and variance σ^2, then we can standardize $N_{\mu,\sigma}$ to get the random variable

$$Z = \frac{N_{\mu,\sigma} - \mu}{\sigma}$$

which has mean 0 and variance 1. To compute the distribution of Z, we have

$$\mathbb{P}(Z \le t) = \mathbb{P}(\mathcal{N}_{\mu,\sigma} \le \sigma t + \mu) = \frac{1}{\sqrt{2\pi\sigma^2}} \int_{-\infty}^{\sigma t + \mu} e^{-\frac{(x-\mu)^2}{2\sigma^2}} dx$$

The substitution $y = (x - \mu)/\sigma$ gives

$$\mathbb{P}(Z \le t) = \frac{1}{\sqrt{2\pi}} \int_{-\infty}^{t} e^{-\frac{y^2}{2}} dx$$

and so Z is also a normal random variable, called the **standard normal random variable**.

Theorem 9.7 *If $\mathcal{N}_{\mu,\sigma}$ is a normal random variable with mean μ and variance σ^2 then*

$$Z = \frac{\mathcal{N}_{\mu,\sigma} - \mu}{\sigma}$$

is a standard normal random variable. Conversely, if Z is a standard normal random variable, then

$$\mathcal{N}_{\tau,\sigma} = \sigma \mathcal{N}_{0,1} + \mu$$

is a normal random variable with mean μ and variance σ^2. $\square$

A distribution related to the normal distribution that we will have use for is the *lognormal* distribution. If a random variable X has the property that its logarithm $\log X$ is normally distributed, then the random variable X is said to have a **lognormal distribution**. Note that X is lognormal if its logarithm is normal, *not* if it is the logarithm of a normal random variable. In other words, lognormal means "log *is* normal" not "log *of* normal." Proof of the following is left as an exercise.

Theorem 9.8 *If X is lognormally distributed; that is, if $Y = \log X$ is normal with mean a and variance b^2, then*

$$\mathcal{E}(X) = \mathcal{E}(e^Y) = e^{a + \frac{1}{2}b^2}$$
$$\mathrm{Var}(X) = \mathrm{Var}(e^Y) = e^{2a + b^2}(e^{b^2} - 1)$$

$\square$

Convergence in Distribution

You may be familiar with the notion of pointwise convergence of a sequence of functions. In any case, here is the definition.

Definition *Let (f_n) be a sequence of functions from $\mathbb{R}$ to $\mathbb{R}$ and let f be another such function. Then (f_n) **converges pointwise** to f if for each real number x, the sequence of real numbers $(f_n(x))$ converges to the real number $f(x)$.* $\square$

Now consider a sequence (X_n) of random variables and let X be a random variable. There are several useful ways in which the notion of *convergence* of the sequence (X_n) to X can be defined. However, we are interested in only one particular form of convergence.

Definition *Let (X_n) be a sequence of random variables, where we allow the possibility that each random variable may be defined on a different probability space $(\Omega_n, \mathbb{P}_n)$. Let X be a random variable on a probability space $(\Omega, \mathbb{P})$. Then (X_n)* **converges in distribution** *to X, written*

$$X_n \xrightarrow{dist} X$$

if the distribution functions (F_{X_n}) converge pointwise to the distribution function F_X at all points where F_X is continuous. Thus, if F_X is continuous at s, then

$$\lim_{n \to \infty} F_{X_n}(s) = F_X(s)$$

that is,

$$\lim_{n \to \infty} \mathbb{P}_n(X_n \le s) = \mathbb{P}(X \le s)$$

Convergence in distribution is also called **weak convergence.** $\square$

It can be shown that if X_n converges in distribution to X, then $f(X_n)$ converges in distribution to $f(X)$ for any continuous function f. Also, weak convergence can be characterized through expected values, as described in the following theorem.

Theorem 9.9 *Let (X_n) be a sequence of random variables, where X_n is defined on $(\Omega_n, \mathbb{P}_n)$. Let X be a random variable defined on $(\Omega, \mathbb{P})$.*
1)   We have

$$X_n \xrightarrow{dist} X \quad \Leftrightarrow \quad \mathcal{E}_{\mathbb{P}_n}(g(X_n)) \to \mathcal{E}_{\mathbb{P}}(g(X))$$

for all bounded continuous functions $g \colon \mathbb{R} \to \mathbb{R}$. In particular,

$$X_n \xrightarrow{dist} X \quad \Rightarrow \quad \mathcal{E}_{\mathbb{P}_n}(X_n) \to \mathcal{E}_{\mathbb{P}}(X)$$

2)   For all continuous functions $f \colon \mathbb{R} \to \mathbb{R}$,

$$X_n \xrightarrow{dist} X \quad \Rightarrow \quad f(X_n) \xrightarrow{dist} f(X)$$ $\square$

We will also need the following result, whose proof we also omit.

Theorem 9.10 *Let (X_n) be a sequence of random variables with*

$$X_n \xrightarrow{dist} X$$

where X is a random variable whose distribution function is continuous. If (a_n) and (b_n) are sequences of real numbers for which

$$a_n \to a \quad and \quad b_n \to b$$

then

$$a_n X_n + b_n \xrightarrow{dist} aX + b$$

In particular, if $a \neq 0$ and $X_n \xrightarrow{dist} \mathcal{N}_{0,1}$ where $\mathcal{N}_{0,1}$ is a standard normal random variable, then

$$a_n X_n + b_n \xrightarrow{dist} \mathcal{N}_{a,b}$$

where $\mathcal{N}_{a,b}$ is a normal random variable with mean a and variance b^2. $\square$

The Central Limit Theorem

The Central Limit Theorem is the most famous theorem in probability and with good reason. Actually, there are several versions of the Central Limit Theorem. We will state the most commonly seen version first and later discuss a different version that we will use in the next chapter.

Speaking intuitively, the distribution function of a random variable X describes its probabilistic behavior. More precisely, if X and Y are random variables with the same distribution function then

$$\mathbb{P}(a \leq X \leq b) = \mathbb{P}(a \leq Y \leq b)$$

for all real numbers a and b. Note that the *functions* X and Y need not be the same. In fact, they need not even be defined on the same sample space. When two or more random variables have the same distribution function, they are said to be **identically distributed**.

Informally speaking, the Central Limit Theorem says that if

$$S_n = X_1 + \cdots + X_n$$

is the sum of n independent random variables X_i that are identically distributed with finite mean μ and finite variance σ^2 and if we standardize S_n to

$$S_n^* = \frac{S_n - n\mu}{\sqrt{n}\sigma}$$

then the distribution function of S_n^* approximates the standard normal distribution *regardless of the type of distribution of the original random* *variables X_i.* Moreover, the approximation gets better and better as n gets larger and larger.

Thus, the process of summing and standardizing "washes out" the original probabilistic characteristics of the individual random variables S_n and replaces them with the characteristics of the standard normal random variable!

Here is a formal statement of the Central Limit Theorem.

Theorem 9.11 (Central Limit Theorem) *Let $X_1, X_2, \ldots$ be a sequence of independent, identically distributed random variables with finite mean μ and finite variance $\sigma^2 > 0$. If*

$$S_n = \sum_{i=1}^{n} X_i$$

then the standardized random variables

$$S_n^* = \frac{S_n - n\mu}{\sqrt{n}\sigma}$$

converge in distribution to a standard normal random variable $N_{0,1}$; that is,

$$\lim_{n \to \infty} F_{S_n^*}(t) = \phi_{0,1}(t)$$

and so

$$P(S_n^* < t) \approx \frac{1}{\sqrt{2\pi}} \int_{-\infty}^{t} e^{-\frac{x^2}{2}} dx$$

where the error in the approximation tends to 0 as n tends to ∞. □

As you might expect, the proof of the Central Limit Theorem is a bit involved and we will not go into it in this book.

As mentioned earlier, we need a different version of the Central Limit Theorem for our work on the Black–Scholes formula, since we must consider not just a simple sequence of random variables but a *triangular array* of random variables

$$
\begin{array}{cccc}
B_{1,1} & & & \\
B_{2,1} & B_{2,2} & & \\
B_{3,1} & B_{3,2} & B_{3,3} & \\
\vdots & \vdots & \vdots & \ddots
\end{array}
$$

In each row, the random variables are independent, identically distributed standard Bernoulli random variables. Thus, $B_{n,i}$ is a Bernoulli random variable with

$$\mathbb{P}\left(B_{n,i} = \frac{q_n}{\sqrt{p_n q_n}}\right) = p_n \quad \text{and} \quad \mathbb{P}\left(B_{n,i} = \frac{-p_n}{\sqrt{p_n q_n}}\right) = q_n$$

where $q_n = 1 - p_n$ (see Theorem 3.17). However, random variables from *different* rows need not be independent, nor are they necessarily identically distributed. In fact, they need not even be defined on the same probability space.

The version of the Central Limit Theorem that covers this situation says that under a certain condition, the adjusted row sums

$$S_1 = B_{1,1}$$
$$S_2 = \frac{1}{\sqrt{2}}(B_{2,1} + B_{2,2})$$
$$S_3 = \frac{1}{\sqrt{3}}(B_{3,1} + B_{3,2} + B_{3,3})$$
$$\vdots$$

converge in distribution to the standard normal distribution $\phi_{0,1}$. Without going into details, it is sufficient to assume that there is a p satisfying $0 < p < 1$ for which

$$p_n \to p$$

Then q_n converges to $q = 1 - p$, which also satisfies $0 < q < 1$.

Theorem 9.12 (Central Limit Theorem) *Consider a triangular array of random variables*

$$\begin{array}{llll} B_{1,1} & & & \\ B_{2,1} & B_{2,2} & & \\ B_{3,1} & B_{3,2} & B_{3,3} & \\ \vdots & \vdots & \vdots & \ddots \end{array}$$

where for each row, the random variables $B_{n,i}$ are independent, identically distributed standard Bernoulli random variables with

$$\mathbb{P}\left(B_{n,i} = \frac{q_n}{\sqrt{p_n q_n}}\right) = p_n \quad \text{and} \quad \mathbb{P}\left(B_{n,i} = \frac{-p_n}{\sqrt{p_n q_n}}\right) = q_n$$

However, the random variables in different rows need not be independent or identically distributed, or even defined on the same probability space. Suppose also that $p_n \to p$, where $0 < p < 1$. Then the random variables

$$S_n = \frac{1}{\sqrt{n}}\sum_{i=1}^{n} B_{n,i}$$

converge in distribution to a standard normal random variable. $\square$

As mentioned earlier, we will use this theorem in the next chapter to help derive the Black–Scholes option pricing formula.

Exercises

1. Let $f(t)$ be a *piecewise linear* probability density function with the following properties: $f(t) = 0$ for $t \leq 0$ and $t \geq 2$, $f(1) = a$. Sketch the graph and find a. Sketch the corresponding distribution function.

2. Let X have a distribution function F given by

$$F(t) = \begin{cases} 0 & t < 0 \\ \frac{1}{2}t & 0 \leq t \leq 2 \\ 1 & t > 2 \end{cases}$$

   Let $Y = X^2$. Find
   a) $\mathbb{P}(0 \leq X \leq 1)$
   b) $\mathbb{P}(1 \leq X \leq 3)$
   c) $\mathbb{P}(Y \leq X)$
   d) $\mathbb{P}(X + Y \leq \frac{3}{4})$
   e) the distribution function of the random variable $\sqrt{X}$

3. Let $\Omega = \{\omega_1, \omega_2, \omega_3\}$ and let $\mathbb{P}$ be the uniform probability measure on Ω; that is, $\mathbb{P}(\omega_i) = 1/3$ for $i = 1, 2, 3$. Consider the following random variables

$$X(\omega_1) = 1, X(\omega_2) = 2, X(\omega_3) = 3$$
$$Y(\omega_1) = 2, Y(\omega_2) = 3, Y(\omega_3) = 1$$
$$Z(\omega_1) = 3, Z(\omega_2) = 1, Z(\omega_3) = 2$$

   Are these functions the same? What about their distribution functions?

4. Show that a σ-algebra is closed under countable intersections.

5. Show directly that all rays are Borel sets.

6. Show directly that all closed intervals are Borel sets.

7. Prove that

$$\mathbb{P}(A \cup B) + \mathbb{P}(A \cap B) = \mathbb{P}(A) + \mathbb{P}(B)$$

   for any events A and B. This is called the **Principle of Inclusion-Exclusion** (for two events).

8. Prove that a probability measure is **subadditive**; that is,

$$\mathbb{P}(A \cup B) \leq \mathbb{P}(A) + \mathbb{P}(B)$$

   for any events A and B.

9. Find and graph the uniform distribution function on the interval $[a, b]$.

10. Show that the most general Bernoulli random variable B with mean 0 and variance 1 is given by

$$\mathbb{P}(B = \frac{q}{\sqrt{pq}}) = p$$

$$\mathbb{P}(B = \frac{-p}{\sqrt{pq}}) = q$$

where $q = 1 - p$.

11. Fill in the details to show that the normal distribution has mean μ.

12. Compute the variance of the normal distribution using the integral formula in the text.

13. Prove that if $A_1 \subseteq A_2 \subseteq \cdots$ is an increasing sequence of events, each contained in the next event, then $\lim_{i \to \infty} \mathbb{P}(A_i)$ exists and

$$\lim_{i \to \infty} \mathbb{P}(A_i) = \mathbb{P}\left(\bigcup_{i=1}^{\infty} A_i\right)$$

14. Let X be a random variable on $\mathbb{R}$. Prove that the function

$$f(t) = \mathbb{P}(X \le t)$$

is a probability distribution function. *Hint*: Make heavy use of monotone continuity.

15. For a probability measure $\mathbb{P}$ with distribution function F verify that
    a) $\mathbb{P}((a, b]) = F(b) - F(a)$
    b) $\mathbb{P}((a, b)) = F(b-) - F(a)$
    c) $\mathbb{P}([a, b)) = F(b-) - F(a-)$
    d) $\mathbb{P}([a, b]) = F(b) - F(a-)$
    where the negative sign means limit from below.

16. If $X_n \xrightarrow{\text{dist}} X$ show that for any real numbers $a \ne 0$ and b

$$aX_n + b \xrightarrow{\text{dist}} aX + b$$

17. Use part 1) of Theorem 9.9 to prove part 2).

18. Prove that if $\mathbb{P}$ and $\mathbb{Q}$ are probability measures on $\mathbb{R}$ and if

$$\mathbb{P}((-\infty, t]) = \mathbb{Q}((-\infty, t])$$

for all $t \in \mathbb{R}$, then $\mathbb{P}$ and $\mathbb{Q}$ also agree on all open intervals.

Chapter 10
The Black–Scholes Option Pricing Formula

The models that we have been studying are *discrete-time* models, because changes take place only at discrete points in time. On the other hand, in *continuous-time* models, changes can take place (at least theoretically) at any real time during the life of the model.

The most famous continuous-time derivative pricing model culminates in the Black–Scholes option pricing formula, which gives the price of a European put or call based on five quantities:

- the *initial price* of the underlying stock, which is known,
- the *strike price* of the option, which is known,
- the *time to expiration*, which is known,
- the *risk-free rate* during the lifetime of the option, which is assumed to be constant and can only be estimated,
- the *volatility* of the stock price, a constant that provides a measure of the fluctuation in the stock's price and thus is a measure of the risk involved in the stock. This quantity can only be estimated as well.

Our goal in this chapter is to describe this continuous-time model and to derive the Black–Scholes option pricing formula. We will derive the continuous-time model as a limiting case of the binomial model.

Stock Prices and Brownian Motion

In 1827, just 35 years after the New York Stock Exchange was founded, an English botanist named Robert Brown studied the motion of small pollen grains immersed in a liquid medium. Brown wrote that pollen grains exhibited a "continuous swarming motion" when viewed under the microscope.

The first mathematical explanation of this phenomenon was given by Thorvald N. Thiele in 1880 in a paper on the method of least squares. In 1905, Albert Einstein showed that this swarming motion, which is now called **Brownian motion**, could be explained as the consequence of the continual bombardment of the particle by the molecules of the liquid. A formal mathematical description of Brownian motion and its properties was first given by the mathematician Norbert Wiener beginning in 1918.

It is especially interesting to note that the phenomenon that is now known as Brownian motion was used by the French mathematician Louis Bachelier to model the movement of stock prices for his doctoral dissertation in 1900.

Brownian Motion

Let us give a formal definition of Brownian motion. We have defined a finite stochastic process as a sequence $X_1, \ldots, X_N$ of random variables defined on a sample space Ω. A **continuous stochastic process** on an interval $I \subseteq \mathbb{R}$ of the real line is a collection $\{X_t \mid t \in I\}$ of random variables on Ω indexed by a variable t that ranges over the interval I. We will generally take I to be the interval $[0, \infty)$. The variable t often represents time and so X_t is the value of the process at time t. If $\{X_t \mid t \in I\}$ is a stochastic process, then the random variable

$$X_t - X_s$$

where $s < t$ is called an **increment** of the process.

Definition *A continuous stochastic process* $\mathcal{W} = \{W_t \mid t \geq 0\}$ *is a* **Brownian motion process** *with* **volatility** σ *if*
1) *$W_0 = 0$*
2) *Each increment $W_t - W_s$ is normally distributed with mean 0 and variance $\sigma^2(t - s)$. In particular, since $W_0 = 0$, each W_t is normally distributed with mean 0 and variance $\sigma^2 t$.*
3) *$\mathcal{W}$ has* **independent increments***; that is, for any times $t_1 \leq t_2 \leq \cdots \leq t_n$, the nonoverlapping increments*

$$W_{t_2} - W_{t_1}, W_{t_3} - W_{t_2}, \ldots, W_{t_n} - W_{t_{n-1}}$$

are independent. $\square$

Brownian Motion with Drift

It is also possible to define Brownian motion with drift. Drift takes the form of a constant multiple of time.

Definition *A stochastic process of the form*

$$\mathcal{W} = \{\mu t + W_t \mid t \geq 0\}$$

where μ is a constant and $\{W_t\}$ is Brownian motion with volatility σ is called **Brownian motion** *with* **drift** μ *and* **volatility** σ.$\square$

When we use the phrase *Brownian motion*, we mean Brownian motion with drift (which may be 0).

Standard Brownian Motion

A Brownian motion process $\{Z_t \mid t \geq 0\}$ with drift $\mu = 0$ and volatility $\sigma = 1$ is called **standard Brownian motion**. In this case Z_t has mean 0 and variance t. If $\{W_t \mid t \geq 0\}$ is Brownian motion with drift μ and variance σ^2, then we can write

$$W_t = \mu t + \sigma Z_t$$

where $\{Z_t \mid t \geq 0\}$ is standard Brownian motion.

Sample Paths

Figure 10.1 shows three simulated *sample paths* for Brownian motion with drift $\mu = 0.08$ and volatility $\sigma = 0.20$ on the interval $[0, 1]$. The straight line shows the drift.

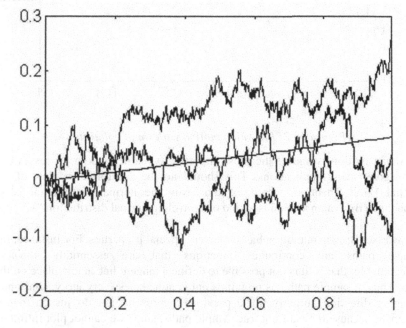

Figure 10.1: Brownian motion sample paths: $\mu = 0.08, \sigma = 0.20$

More specifically, if we fix an outcome $\omega \in \Omega$, then we can define a function

$$t \to W_t(\omega)$$

The graph of this function is called a **sample path**.

Figure 10.2 shows a single sample path for a Brownian motion process with the same mean $\mu = 0.08$ as the previous case but with volatility $\sigma = 0.05$. As you can see, volatility is aptly named.

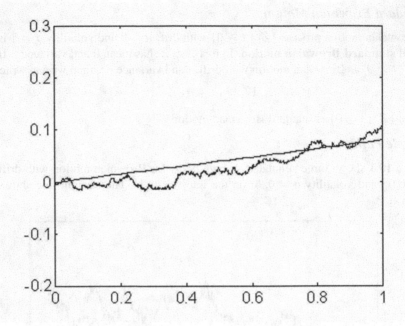

Figure 10.2: Brownian motion with small volatility

Brownian motion is one of the most important types of random processes and has a great many applications. This should not be a surprise in view of the Central Limit Theorem, which explains why the normal distribution is so important. (Brownian motion is a kind of "traveling normal distribution.")

In any case, Brownian motion has some very special properties. For instance, the sample paths are continuous functions that are essentially *nowhere* differentiable; that is, it is not possible to define a tangent line at any place on the path! Thus, a sample path has no jumps but is nonetheless very jerky, constantly changing direction abruptly. (The previous figures do not do justice to this statement because they are not true sample paths, since we cannot plot infinitely many values.)

Geometric Brownian Motion and Stock Prices

How does Brownian motion with drift relate to stock prices? One *possibility* is to think of a stock's price as a "particle" that is subject to constant bombardment by "smaller particles" in the form of individual stock trades, various changes in the economy and other events. Indeed, the normal distribution seems like a reasonable choice to model a stock's price if we think of the vicissitudes of that price as being the result of a large number of more or less independent (and similarly distributed) factors.

However, there are some obvious problems with the Brownian motion viewpoint for stock prices themselves. First, a Brownian motion process can be negative whereas stock prices are never negative. Second, in a Brownian motion process, the expected change

$$\mathcal{E}(W_t - W_s) = \mu(t - s)$$

in the value of $\mathcal{W}$ depends only on the time change $t - s$ and not on the value W_s, for example. But the expected change in a stock price does seem to depend on the stock's starting price W_s. For instance, an expected price change of \$1 might be quite reasonable if $W_s = \$200$ but quite unreasonable if $W_s = \$0.02$.

To resolve these issues, it would seem to make more sense to model the *instantaneous rate of return* of the stock price as a Brownian motion process, for this quantity seems more reasonably independent of the initial price. For example, to say that a stock's rate of return is 10% is to say that the price may grow by a *factor* of $e^{0.1}$, regardless of its initial value.

This viewpoint can be handled by assuming that the stock price S_t at time t is given by

$$S_t = S_0 e^{H_t}$$

where S_0 is the initial price and H_t is a Brownian motion process. The exponent H_t represents a *continuously compounded* rate of return of the stock price over the period of time $[0, t]$. Note also that H_t satisfies

$$H_t = \log\left(\frac{S_t}{S_0}\right)$$

and so we will refer to it as the **logarithmic growth** of the stock price.

Definition *A stochastic process of the form*

$$\{e^{W_t} \mid t \geq 0\}$$

where $\{W_t \mid t \geq 0\}$ is Brownian motion is called **geometric Brownian motion.** $\square$

Figure 10.3 shows a simulated sample path from a geometric Brownian motion process with the same drift ($\mu = 0.08$) as before but with an unnaturally large volatility in order to demonstrate the exponential nature of the growth.

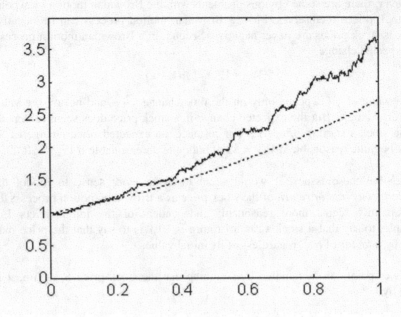

Figure 10.3: Geometric Brownian motion

If we assume that H_t follows a Brownian motion process with positive drift, then we can write

$$H_t = \log\left(\frac{S_t}{S_0}\right) = \mu t + \sigma W_t$$

where $\{W_t\}$ is standard Brownian motion. Therefore, H_t has a normal distribution with

$$\mathcal{E}(H_t) = \mu t \quad \text{and} \quad \text{Var}(H_t) = \sigma^2 t$$

As we have seen, if a random variable X has the property that its logarithm is normally distributed, then the random variable X is said to have a **lognormal distribution**. Accordingly, S_t/S_0 is lognormally distributed. Figure 10.4 shows a typical lognormal density function.

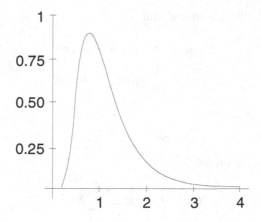

Figure 10.4: A lognormal density function

According to Theorem 9.8,

$$\mathcal{E}(S_t) = S_0 e^{(\mu + \frac{1}{2}\sigma^2)t}$$
$$\text{Var}(S_t) = (S_0 e^{(\mu + \frac{1}{2}\sigma^2)t})^2 (e^{\sigma^2 t} - 1)$$

This value of $\mathcal{E}(S_t)$ tells us that the expected stock price depends not only on the drift μ of H_t but also on the volatility σ.

The Binomial Model in the Limit: Brownian Motion

In an earlier chapter, we looked at the binomial model from the point of view of the logarithmic growth in the stock price. Let us recall the pertinent results.

Recap of the Binomial Model

First, we specify the model times as

$$0 = t_0 < t_1 < \cdots < t_T = t$$

Thus, the lifetime of the model is $[0, t]$. We are using t here instead of L because we want to think of t as a variable. We also want to emphasize the role of the number T of time intervals in the model, since our plan is to let T approach infinity. Now let us recall Theorem 6.4.

Theorem 10.1 *Let* $\mathbb{M}_T$ *be a binomial model with the following parameters:*
1) *Lifetime* $t \geq 0$.
2) *T time increments, each of length* $\Delta t = t/T$.
3) *Up-tick factor u and down-tick factor d, where* $0 < d \leq 1 \leq u$.
4) *A probability of up-tick $p \in (0, 1)$ and down-tick $q = 1 - p$.*

5) *Drift and volatility associated with p,*

$$\mu = \frac{1}{\Delta t}(p\log u + q\log d)$$

$$\sigma^2 = \frac{1}{\Delta t}pq(\log u - \log d)^2$$

Then the stock price is given by

$$S_T = S_0 e^{H_T}$$

where the logarithmic growth is

$$H_T = \mu t + \sigma\sqrt{\Delta t}\sum_{i=1}^{T}X_{T,i}$$

with **deterministic part** μt *and* **random part**

$$Q_T = \sigma\sqrt{\Delta t}\sum_{i=1}^{T}X_{T,i}$$

The random variables $X_{T,i}$ are independent standard Bernoulli random variables defined by

$$X_{T,i} = \begin{cases} \frac{q}{\sqrt{pq}} & \text{with probability } p \\ \frac{-p}{\sqrt{pq}} & \text{with probability } q \end{cases} \qquad \square$$

Note that the parameters p, q, u, d, μ and σ depend on T and so we may include a subscript T when necessary to avoid confusion.

Taking the Limit as $\Delta t \to 0$

Let us now see what happens to the binomial models $\mathbb{M}_T$ as $T \to \infty$ or equivalently, as $\Delta t \to 0$. We want to apply the Central Limit Theorem to the triangular array of Bernoulli random variables that represent the random portion of the stock price:

$$\begin{array}{llll} \text{Model } \mathbb{M}_1: & X_{1,1} \\ \text{Model } \mathbb{M}_2: & X_{2,1} & X_{2,2} \\ \text{Model } \mathbb{M}_3: & X_{3,1} & X_{3,2} & X_{3,3} \\ & \vdots & \vdots & \vdots & \ddots \end{array}$$

For each model $\mathbb{M}_k$ with probability space $(\Omega_T, \mathbb{P}_T)$, the random variables $X_{k,i}$ are independent and identically distributed, with

$$\mathbb{P}_T\left(X_{T,i} = \frac{1 - p_T}{\sqrt{p_T(1 - p_T)}}\right) = p_T$$

$$\mathbb{P}_T\left(X_{T,i} = \frac{-p_T}{\sqrt{p_T(1 - p_T)}}\right) = 1 - p_T$$

The Central Limit Theorem in the form of Theorem 9.12 applies to the triangular array above provided that the probabilities p_T satisfy

Requirement 1: $p_T \to p$ for some $p \in (0, 1)$

Assuming that this requirement is met, the Central Limit Theorem implies that the adjusted sums

$$\frac{1}{\sqrt{T}}\sum_{i=1}^{T} X_{T,i} = \frac{1}{\sigma_T\sqrt{t}}Q_T$$

converge in distribution to a standard normal random variable; that is, for any real number s,

$$\lim_{T\to\infty}\mathbb{P}_T\left(\frac{Q_T}{\sigma_T\sqrt{t}} < s\right) = \phi_{0,1}(s)$$

Thus, we may write

$$\frac{Q_T}{\sigma_T\sqrt{t}} \xrightarrow{\text{dist}} Z_t$$

where Z_t is a standard normal random variable.

For the logarithmic growth H_T, Theorem 9.10 implies that if the following holds:

Requirement 2: $\mu_T \to \mu, \sigma_T \to \sigma$ for real numbers μ and $\sigma \neq 0$

then

$$Q_T \xrightarrow{\text{dist}} \sigma_T\sqrt{t}Z_t$$

and so

$$H_T = \mu_T t + Q_T \xrightarrow{\text{dist}} H_t := \mu t + \sigma\sqrt{t}Z_t$$

As to the stock price itself, since S_T is a continuous function of H_T,

$$S_T = S_0 e^{H_T}$$

and so Theorem 9.9 implies that

$$S_T \xrightarrow{\text{dist}} S_0 e^{H_t} = S_0 e^{\mu t + \sigma \sqrt{t} Z_t}$$

Note that

$$\mathcal{E}(\sqrt{t} Z_t) = 0 \quad \text{and} \quad \text{Var}(\sqrt{t} Z_t) = t$$

and

$$\mathcal{E}(H_t) = 0 \quad \text{and} \quad \text{Var}(H_t) = \sigma^2 t$$

For convenience, we make the following definition.

Definition *A family of binomial models* $\mathbb{M}_T$ *for* $T > 0$ *is* **stable** *if*
1) $p_T \to p$ *for some* $p \in (0, 1)$.
2) $\mu_T \to \mu, \sigma_T \to \sigma$ *for real numbers* μ *and* $\sigma \neq 0$. $\square$

Variable Lifetimes

Although we have derived these formulas for a fixed total lifetime t, as mentioned earlier, we want to think of t as a variable and so we get one binomial model $\mathbb{M}_{t,T}$ for each lifetime $t > 0$ and interval count T. Unfortunately, our derivation does not expose the relationship between the limiting random variables $\sqrt{t} Z_t$, as t ranges over all nonnegative real numbers.

We will not go into this issue formally. However, we can make a few informal observations. First, it should be intuitively clear that our model is *translation invariant* in the sense that we obtain essentially the same model over any two lifetimes $[s_1, t_1]$ and $[s_2, t_2]$ of the same length. Second, because we assume that nonoverlapping increments are independent, we should be able to piece together models from disjoint contiguous intervals into a model for one large interval. These facts imply that the stochastic process

$$\{\sqrt{t} Z_t \mid t \geq 0\}$$

is actually standard Brownian motion. Hence,

$$\{H_t \mid t \geq 0\}$$

is a Brownian motion process with drift μ and volatility σ. Finally, the stock price process itself

$$\{S_t \mid t \geq 0\}$$

is a geometric Brownian motion process with drift μ and volatility σ^2.

We can now summarize our knowledge of the behavior of the stock price in the limiting case of the binomial model.

Theorem 10.2 *Let* $\mathbb{M}_{t,T}$ *be a binomial model with the following parameters:*

1) *Lifetime* $t \geq 0$.
2) *T time increments, each of length $\Delta t = t/T$.*
3) *Up-tick factor u_T and down-tick factor d_T, where $0 < d_T \leq 1 \leq u_T$.*
4) *Probability of up-tick $p_T \in (0,1)$ and down-tick $q_T = 1 - p_T$.*
5) *Drift and volatility*

$$\mu_T = \frac{1}{\Delta t}(p_T \log u_T + q_T \log d_T)$$

$$\sigma_T^2 = \frac{1}{\Delta t} p_T q_T (\log u_T - \log d_T)^2$$

Let $S_{t,T}$ be the final stock price and let

$$H_{t,T} = \log\left(\frac{S_{t,T}}{S_0}\right)$$

be the logarithmic price growth. If the family $\{\mathbb{M}_{t,T} \mid T > 0\}$ is stable; that is, if

$$p_T \to p$$

for some $p \in (0,1)$ and

$$\mu_T \to \mu, \sigma_T \to \sigma$$

for some real numbers μ and $\sigma \neq 0$, then

$$H_{t,T} \xrightarrow{dist} H_t = \mu t + \sigma \sqrt{t} Z_t$$

and

$$S_{t,T} \xrightarrow{dist} S_t = S_0 e^{\mu t + \sigma \sqrt{t} Z_t}$$

Moreover, the process

$$\{\sqrt{t} Z_t \mid t \geq 0\}$$

is standard Brownian motion. Hence, the logarithmic growth and stock price processes

$$\{H_t \mid t \geq 0\} \quad and \quad \{S_t \mid t \geq 0\}$$

are Brownian and geometric Brownian, respectively, with drift μ and volatility σ. Also, the stock price growth S_t/S_0 is lognormally distributed with

$$\mathcal{E}(S_t) = S_0 e^{(\mu + \frac{1}{2}\sigma^2)t}$$

$$\text{Var}(S_t) = (S_0 e^{(\mu + \frac{1}{2}\sigma^2)t})^2 (e^{\sigma^2 t} - 1) \qquad \square$$

The Natural Binomial Model

In the general binary model described above, the up-tick probability p is arbitrary (but different from 0 or 1). This probability may or may not be related to the martingale measure up-tick π. If the probability p is determined by empirical means (or by assumptions) based on economic data, we refer to it as the **natural up-tick probability** and denote it by ν. We also refer to the binomial model in which $p = \nu$ as the **natural binomial model**. *This is the model under which we want to price alternatives.*

It is customary to make the following assumptions about natural binomial models.

Assumption 1
The natural probability ν of an up-tick in the stock price satisfies $0 < \nu < 1$ and does not depend on the number T of intervals.

Assumption 2
The resulting natural drift and volatility

$$\mu = \frac{1}{\Delta t}(\nu \log u_T + (1 - \nu) \log d_T)$$

$$\sigma^2 = \frac{1}{\Delta t}\nu(1 - \nu)(\log u_T - \log d_T)^2$$

called the **natural drift** and **natural volatility**, respectively, do not depend on the value of T (or Δt). (These parameters are more commonly called the **instantaneous drift** and **instantaneous volatility**.)

Now, the assumption that μ and σ are *constant* with respect to T amounts to specifying a relationship among u_T, d_T and ν, which is obtained simply by solving the equations above to get

$$\log u_T = \mu \Delta t + \sigma \sqrt{\Delta t} \frac{1 - \nu}{\sqrt{\nu(1 - \nu)}}$$

$$\log d_T = \mu \Delta t - \sigma \sqrt{\Delta t} \frac{\nu}{\sqrt{\nu(1 - \nu)}}$$

Thus, the martingale measure up-tick probability

$$\pi_T = \frac{e^{rT} - d_T}{u_T - d_T}$$

can also be expressed directly in terms of the natural probability ν and the natural drift μ and volatility σ. This expression can be used to show that the martingale measure up-tick probability π_T approaches the natural up-tick probability ν as T approaches ∞.

Theorem 10.3 *Let* $\mathbb{M}_T$ *be the natural binomial model with natural drift* μ *and volatility* σ.

1)   The martingale measure up-tick probability is given by

$$\pi = \frac{e^{(r-\mu)\Delta t + \frac{\nu}{\sqrt{\nu(1-\nu)}}\sigma\sqrt{\Delta t}} - 1}{e^{\frac{1}{\sqrt{\nu(1-\nu)}}\sigma\sqrt{\Delta t}} - 1}$$

2)   The martingale probability $\pi = \pi_T$ *approaches the natural probability* ν *as* $T \to \infty$ *or equivalently as* $\Delta t \to 0$; *that is,*

$$\lim_{T\to\infty} \pi_T = \nu$$

Proof. For part 1), the previous equations for $\log u_T$ and $\log d_T$ give

$$u_T = e^{\mu\Delta t + \sigma\sqrt{\Delta t}\frac{1-\nu}{\sqrt{\nu(1-\nu)}}} \quad \text{and} \quad d_T = e^{\mu\Delta t - \sigma\sqrt{\Delta t}\frac{\nu}{\sqrt{\nu(1-\nu)}}}$$

Substituting into the expression for π_T gives the desired formula. Part 2) is a simple application of l'Hopital's rule to evaluate the limit

$$\lim_{T\to\infty} \pi_T = \lim_{\Delta t\to 0} \frac{e^{(r-\mu)\Delta t + \frac{\nu}{\sqrt{\nu(1-\nu)}}\sigma\sqrt{\Delta t}} - 1}{e^{\frac{1}{\sqrt{\nu(1-\nu)}}\sigma\sqrt{\Delta t}} - 1}$$

We leave this to the reader. $\square$

Of course, the assumptions that ν, μ and σ are constant and $0 < \nu < 1$ imply that the natural binomial models are stable and so Theorem 10.2 implies that

$$H_T \xrightarrow{\text{dist}} H_t = \mu t + \sigma\sqrt{t}Z_t$$

and

$$S_T \xrightarrow{\text{dist}} S_t = S_0 e^{\mu t + \sigma\sqrt{t}Z_t}$$

where

$$\{\sqrt{t}Z_t \mid t \ge 0\}$$

is standard Brownian motion. Also, the logarithmic growth and stock price

$$\{H_t \mid t \ge 0\} \quad \text{and} \quad \{S_t \mid t \ge 0\}$$

are Brownian and geometric Brownian, respectively, with drift μ and volatility σ. Also, the stock price growth S_t/S_0 is lognormally distributed with

$$\mathcal{E}(S_t) = S_0 e^{(\mu+\frac{1}{2}\sigma^2)t}$$
$$\text{Var}(S_t) = (S_0 e^{(\mu+\frac{1}{2}\sigma^2)t})^2 (e^{\sigma^2 t} - 1)$$

Estimating the Natural Drift and Volatility

The assumption that the natural up-tick probability, the natural drift and the natural volatility are constant is not very realistic, but is based on practical considerations, since the model would be mathematically intractable without some simplifying assumptions (as is often the case with complex physical phenomena). In fact, this assumption is usually extended into the past; that is, it is assumed that the natural drift and volatility can be estimated by looking at the *past history* of the price for the stock in question.

Specifically, we choose a small value for Δt; for example, Δt may correspond to one day. Then over a large number of these short periods of time, we compute the logarithmic growth factors

$$\log E_i = \log\left(\frac{\text{Stock price at end of period}}{\text{Stock price at start of period}}\right)$$

The average of these sample values is an estimate of $\mu \Delta t$ and the sample variance of these sample values is an estimate of $\sigma^2 \Delta t$. Of course, the more samples we take, the better the estimates will be. Let us consider an example.

Example 10.1 The following portion of an Excel spreadsheet shows closing stock prices over a 10 day period. Here we are taking Δt to be one day; that is, $1/365$ years. The initial price is $50.

Day	Price	Growth	Log Growth	Average	Sample Var
0	50			0.000359354	0.000677264
1	50.95	1.019	0.018821754	**Per Unit**	**Per Unit**
2	49.74	0.976251	-0.024035321	0.131164046	0.247201227
3	49.46	0.994371	-0.005645176		
4	49.83	1.007481	0.00745295		
5	48.7	0.977323	-0.022938182		
6	50.2	1.030801	0.030335997		
7	49.57	0.98745	-0.012629215		
8	51.78	1.044583	0.043618162		
9	52.17	1.007532	0.007503643		
10	50.18	0.961855	-0.038891076		

The growth column contains the quotient of the stock price and the previous stock price. For example, $50.95/50 = 1.019$. The average is simply the sum of the logarithmic growths divided by the number of logarithmic growths. The sample variance is computed using the formula

$$\text{Sample var} = \frac{1}{n-1}\sum_{i=1}^{n}(i\text{th value} - \text{average})^2$$

(The reason for dividing by $n-1$ instead of n has to do with obtaining an unbiased value.) Finally, the "per unit" values are obtained by multiplying the average and variance by $1/\Delta t = 365$. It follows that $\mu \approx 0.13$ and $\sigma^2 \approx 0.25$ on an annual basis (that is, the units of time are measured in years).$\square$

The Martingale Measure Binomial Model

Let us take a peek at our main goal to see what to do next. The payoff for a European put (for example) with strike price K is

$$X = (K - S_T)^+ = (K - S_0 e^{H_T})^+$$

and the absence of arbitrage implies that the initial price of the put must be

$$\mathcal{I}(\text{Put}) = e^{-rt}\mathcal{E}((K - S_0 e^{H_T})^+)$$

where the expected value is taken under the martingale measure of the natural binomial model. Taking limits as T tends to infinity gives

$$P_\infty = \lim_{T\to\infty} \mathcal{I}(\text{Put}) = e^{-rt} \lim_{T\to\infty} \mathcal{E}((K - S_0 e^{H_T})^+)$$

where P_∞ denotes the limiting price random variable. Setting

$$g(x) = (K - S_0 e^x)^+$$

which is bounded and continuous on $\mathbb{R}$ gives

$$P_\infty = e^{-rt} \lim_{T\to\infty} \mathcal{E}(g(H_T))$$

Now Theorem 9.9 says that if H_T converges weakly to a random variable X *under the martingale measure*, then

$$P_\infty = e^{-rt} \lim_{T\to\infty} \mathcal{E}(g(H_T)) = e^{-rt}\mathcal{E}(g(X))$$

Hence, we must find the weak limit of H_T under the martingale measure.

To find this weak limit, we consider the binomial models $\mathbb{M}_{\pi,T}$ formed by taking p_T to be the martingale measure up-tick probability from the *natural model* with natural up-tick probability ν; that is,

$$p_T = \pi_T = \frac{e^{rT} - d_T}{u_T - d_T}$$

Let us refer to this as the **martingale measure binomial model**. To see if these models are stable, we must determine the limits of π_T as well as the martingale measure drift and volatility

$$\mu_{\pi,T} = \frac{1}{\Delta t}(\pi_T \log u_T + (1 - \pi_T)\log d_T)$$

$$\sigma^2_{\pi,T} = \frac{1}{\Delta t}\pi_T(1 - \pi_T)(\log u_T - \log d_T)^2$$

as T approaches ∞. Theorem 10.3 implies that

$$\lim_{T \to \infty} \pi_T = \nu$$

To determine the other limits, we relate the drifts and volatilities of the two models.

Theorem 10.4 *The martingale measure model drift $\mu_{\pi,T}$ and volatility $\sigma_{\pi,T}$ are related to the natural drift μ_ν and volatility σ_ν by*

$$\mu_{\pi,T} = \mu_\nu + \sigma_\nu \frac{1}{\sqrt{\Delta t}} \frac{(\pi_T - \nu)}{\sqrt{\nu(1-\nu)}}$$

$$\sigma_{\pi,T}^2 = \sigma_\nu^2 \frac{\pi_T(1-\pi_T)}{\nu(1-\nu)}$$

Proof. For the drifts, we have

$$\mu_{\pi,T} - \mu_\nu = \frac{1}{\Delta t}(\pi_T - \nu)(\log u - \log d)$$

and since

$$\sigma_\nu = \frac{1}{\sqrt{\Delta t}} \sqrt{\nu(1-\nu)}(\log u - \log d)$$

we have

$$\frac{\mu_{\pi,T} - \mu_\nu}{\sigma_\nu} = \frac{1}{\sqrt{\Delta t}} \frac{\pi_T - \nu}{\sqrt{\nu(1-\nu)}}$$

from which the formula follows. For the volatilities, we have

$$\frac{\sigma_{\pi,T}^2}{\sigma_\nu^2} = \frac{\frac{1}{\Delta t}\pi_T(1-\pi_T)(\log u - \log d)^2}{\frac{1}{\Delta t}\nu(1-\nu)(\log u - \log d)^2} = \frac{\pi_T(1-\pi_T)}{\nu(1-\nu)} \qquad \square$$

Now we can compute the required limits for stability of the martingale measure binomial models.

Theorem 10.5 *The following limits hold:*

$$\lim_{T \to \infty} \mu_{\pi,T} = r - \frac{\sigma_\nu^2}{2} \quad and \quad \lim_{T \to \infty} \sigma_{\pi,T} = \sigma_\nu$$

where r is the risk-free rate.
Proof. Since μ_ν and σ_ν are constant and Theorem 10.4 gives

$$\mu_{\pi,T} = \mu_\nu + \sigma_\nu \frac{1}{\sqrt{\Delta t}} \frac{(\pi - \nu)}{\sqrt{\nu(1-\nu)}}$$

we have

$$\lim_{T \to \infty} \mu_{\pi,T} = \mu_v + \frac{\sigma_v}{\sqrt{\nu(1-\nu)}} \lim_{\Delta t \to 0} \left(\frac{\pi_T - \nu}{\sqrt{\Delta t}} \right)$$

Now we use the formula for π from Theorem 10.3 to evaluate the limit

$$\lim_{\Delta t \to 0} \frac{1}{\sqrt{\Delta t}} \left(\frac{e^{(r-\mu_v)\Delta t + \frac{\nu}{\sqrt{\nu(1-\nu)}}\sigma_v\sqrt{\Delta t}} - 1}{e^{\frac{1}{\sqrt{\nu(1-\nu)}}\sigma_v\sqrt{\Delta t}} - 1} - \nu \right)$$

This is a bit messy. Either l'Hopital's rule and a strong constitution or a symbolic algebra software package gives the limit as

$$\frac{2(r - \mu_v) - \sigma_v^2}{2\sigma_v} \sqrt{\nu(1-\nu)}$$

and so

$$\lim_{T \to \infty} \mu_{\pi,T} = \mu_v + \frac{\sigma_v}{\sqrt{\nu(1-\nu)}} \left[\frac{2(r - \mu_v) - \sigma_v^2}{2\sigma_v} \sqrt{\nu(1-\nu)} \right]$$

$$= r - \frac{\sigma_v^2}{2}$$

as desired. For the limit of the sequence $\sigma_{\pi,T}$, we begin with the formula

$$\sigma_{\pi,T}^2 = \sigma_v^2 \frac{\pi_T(1 - \pi_T)}{\nu(1 - \nu)}$$

from Theorem 10.4. But Theorem 10.3 implies that $\pi_T \to \nu$ and so

$$\lim_{\Delta t \to 0} \sigma_{\pi,T}^2 = \sigma_\nu^2$$

Taking square roots gives the desired result.□

Now Theorem 10.2 will give us the desired weak limit.

Theorem 10.6 *Let $\mathbb{M}_{t,T}$ be a martingale measure binomial model with the following parameters:*
1) *Lifetime $t \geq 0$.*
2) *T time increments, each of length $\Delta t = t/T$.*
3) *Up-tick factor u_T and down-tick factor d_T, where $0 < d_T \leq 1 \leq u_T$.*
4) *Probability of up-tick equal to the martingale measure up-tick probability from the natural model with up-tick probability ν; that is,*

$$p_T = \pi_T = \frac{e^{rT} - d_T}{u_T - d_T}$$

5) Drift and volatility

$$\mu_{\pi,T} = \frac{1}{\Delta t}(\pi_T \log u_T + (1 - \pi_T)\log d_T)$$

$$\sigma_{\pi,T}^2 = \frac{1}{\Delta t}\pi_T(1 - \pi_T)(\log u_T - \log d_T)^2$$

Then

$$\pi_T \to \nu, \quad \mu_{\pi,T} \to r - \frac{\sigma_\nu^2}{2} \quad \text{and} \quad \sigma_{\pi,T} \to \sigma_\nu$$

where r is the risk-free rate and so the model is stable. Hence, the logarithmic growth satisfies

$$H_{t,T} \xrightarrow{dist} H_t = \left(r - \frac{\sigma_\nu^2}{2}\right)t + \sigma_\nu \sqrt{t}Z_t$$

where Z_t is standard normal and

$$S_{t,T} \xrightarrow{dist} S_t = S_0 e^{\left(r - \frac{\sigma_\nu^2}{2}\right)t + \sigma_\nu \sqrt{t}Z_t}$$

where

$$\{\sqrt{t}Z_t \mid t \geq 0\}$$

is standard Brownian motion. The logarithmic growth and stock price processes

$$\{H_t \mid t \geq 0\} \quad \text{and} \quad \{S_t \mid t \geq 0\}$$

are Brownian and geometric Brownian, respectively, with drift μ and volatility σ. Also, the stock price growth S_t/S_0 is lognormally distributed with

$$\mathcal{E}(S_t) = S_0 e^{rt}$$
$$\text{Var}(S_t) = (S_0 e^{rt})^2(e^{\sigma_\nu^2 t} - 1) \qquad \square$$

We will return to our primary goal of alternative pricing under the natural binomial model a bit later. First, we have some other business.

Are the Assumptions Realistic?

We cannot continue without a short pause to comment on whether the assumption that stock prices are lognormally distributed, or equivalently that the logarithmic growth in the stock price is normally distributed, is a realistic one. There has been much statistical work done on this question, resulting in evidence that growth rates exhibit a phenomenon known as *leptokurtosis*, which means two things:

- The probability that the logarithmic growth is near the mean is greater than it would be for a normal distribution (higher peak).

- The probability that the logarithmic growth is far away from the mean is greater than it would be for a normal distribution (fatter tails).

There is other statistical evidence that the assumption of normality is perhaps not realistic. For a further discussion, with references, we refer the reader to Chriss (1997). Of course, this should not necessarily come as a surprise. After all, the assumptions that lead to the formula

$$H_t = \left(r - \frac{\sigma_\nu^2}{2} \right) t + \sigma_\nu \sqrt{t} Z_{\pi, t}$$

are not very realistic. Nevertheless, the Black–Scholes formula, which relies on this normal model, and to which we know turn, is the most widely used formula for option pricing.

The Black–Scholes Option Pricing Formula

We now have the tools necessary to derive the Black–Scholes option pricing formula for a European option. As we have seen, the payoff for a European put (for example) with strike price K is

$$X = (K - S_T)^+ = (K - S_0 e^{H_T})^+$$

and the absence of arbitrage implies that the initial price of the put must be

$$\mathcal{I}(\text{Put}) = e^{-rt} \mathcal{E}((K - S_0 e^{H_T})^+)$$

where the expected value is taken under the martingale measure of the natural binomial model. Taking limits as T tends to infinity gives

$$P_\infty = \lim_{T \to \infty} \mathcal{I}(\text{Put}) = e^{-rt} \lim_{T \to \infty} \mathcal{E}((K - S_0 e^{H_T})^+)$$

where P_∞ denotes the limiting price random variable. But the function

$$g(x) = (K - S_0 e^x)^+$$

is bounded and continuous on $\mathbb{R}$ and since Theorem 10.6 implies that

$$H_T \xrightarrow{\text{dist}} H_t = \left(r - \frac{\sigma_\nu^2}{2} \right) t + \sigma_\nu \sqrt{t} Z_t$$

Theorem 9.9 implies that

$$P_\infty = e^{-rt} \mathcal{E}((K - S_0 e^{H_t})^+)$$

Since H_t is normally distributed with mean and variance

$$a = \left(r - \frac{\sigma_\nu^2}{2} \right) t \quad \text{and} \quad b^2 = \sigma_\nu^2 t$$

respectively, we have

$$P_\infty = \frac{e^{-rt}}{\sqrt{2\pi b^2}} \int_{-\infty}^{\infty} (K - S_0 e^x)^+ e^{-(x-a)^2/2b^2} \, dx$$

where the π under the square root sign is just pi, not the martingale up-tick probability! (Sometimes a π is just a π.)

All we need to do now is evaluate this integral. Making the substitution $y = (x - a)/b$ gives

$$P_\infty = \frac{e^{-rt}}{\sqrt{2\pi}} \int_{-\infty}^{\infty} (K - S_0 e^{a+by})^+ e^{-y^2/2} \, dy$$

Now the integrand is nonzero if and only if

$$K - S_0 e^{a+by} > 0$$

that is, if and only if

$$y < h := \frac{1}{b}\left[\log\left(\frac{K}{S_0}\right) - a\right]$$

and so

$$P_\infty = \frac{e^{-rt}}{\sqrt{2\pi}} \int_{-\infty}^{h} (K - S_0 e^{by+a}) e^{-y^2/2} \, dy$$

Splitting this into two integrals gives

$$P_\infty = \frac{Ke^{-rt}}{\sqrt{2\pi}} \int_{-\infty}^{h} e^{-y^2/2} \, dy - \frac{S_0 e^{-rt}}{\sqrt{2\pi}} \int_{-\infty}^{h} e^{a+by} e^{-y^2/2} \, dy$$

The first of these integrals is

$$\frac{Ke^{-rt}}{\sqrt{2\pi}} \int_{-\infty}^{h} e^{-y^2/2} \, dy = Ke^{-rt} \phi_{0,1}(h)$$

where $\phi_{0,1}(x)$ is the standard normal distribution function. The second integral could use some work

$$\frac{S_0 e^{-rt}}{\sqrt{2\pi}} \int_{-\infty}^{h} e^{a+by} e^{-y^2/2} dy = \frac{S_0 e^{-rt}}{\sqrt{2\pi}} \int_{-\infty}^{h} e^{-(y^2 - 2by - 2a)/2} dy$$

$$= \frac{S_0 e^{-rt}}{\sqrt{2\pi}} \int_{-\infty}^{h} e^{-[(y-b)^2 - b^2 - 2a]/2} dy$$

$$= \frac{S_0 e^{-rt}}{\sqrt{2\pi}} e^{(b^2 + 2a)/2} \int_{-\infty}^{h} e^{-(y-b)^2/2} dy$$

$$= S_0 e^{-rt + b^2/2 + a} \frac{1}{\sqrt{2\pi}} \int_{-\infty}^{h-b} e^{-y^2/2} dy$$

$$= S_0 e^{-rt + b^2/2 + a} \phi_{0,1}(h - b)$$

Thus

$$P_\infty = K e^{-rt} \phi_{0,1}(h) - S_0 e^{-rt + b^2/2 + a} \phi_{0,1}(h - b)$$

Now we have a pleasant surprise with respect to the exponent in the second term, namely,

$$-rt + \frac{1}{2} b^2 + a = -rt + \frac{1}{2} t\sigma_\nu^2 + \frac{t}{2}(2r - \sigma_\nu^2) = 0$$

and so we arrive at our final destination

$$P_\infty = K e^{-rt} \phi_{0,1}(h) - S_0 \phi_{0,1}(h - \sqrt{t}\sigma_\nu)$$

where

$$h = \frac{1}{b}\left[\log\left(\frac{K}{S_0}\right) - a\right] = \frac{1}{\sqrt{t}\sigma_\nu}\left[\log\left(\frac{K}{S_0}\right) - rt + \frac{1}{2} t\sigma_\nu^2\right]$$

This is the famous Black–Scholes formula for the value of a European put.

We can use the put-call option parity formula to get the corresponding price of a European call. Recall that this formula says that the price of a call is given by

$$C = P + S_0 - K e^{-rt}$$

Taking limits as T tends to ∞ gives

$$C_\infty = P_\infty + S_0 - K e^{-rt}$$

$$= K e^{-rt}(\phi_{0,1}(h) - 1) - S_0(\phi_{0,1}(h - \sqrt{t}\sigma_\nu) - 1)$$

Since

$$\phi_{0,1}(-t) = 1 - \phi_{0,1}(t)$$

the price of the call is

$$C_\infty = -Ke^{-rt}\phi_{0,1}(-h) + S_0\phi_{0,1}(\sqrt{t}\sigma_\nu - h)$$

By setting $d_2 = -h$ and $d_1 = d_2 + \sigma\sqrt{t}$ as seems to be commonly done, we get the formulas shown in the following theorem.

Theorem 10.7 (The Black–Scholes Option Pricing Formulas) *For European options with strike price K and expiration time t, we have*

$$C = S_0\phi_{0,1}(d_1) - Ke^{-rt}\phi_{0,1}(d_2)$$
$$P = Ke^{-rt}\phi_{0,1}(-d_2) - S_0\phi_{0,1}(-d_1)$$

where S_0 is the initial price of the underlying stock, σ is the natural volatility, $\phi_{0,1}$ is the standard normal distribution function and

$$d_1 = \frac{1}{\sigma\sqrt{t}}\left[\log\left(\frac{S_0}{K}\right) + t\left(r + \frac{1}{2}\sigma^2\right)\right]$$
$$d_2 = \frac{1}{\sigma\sqrt{t}}\left[\log\left(\frac{S_0}{K}\right) + t\left(r - \frac{1}{2}\sigma^2\right)\right] = d_1 - \sigma\sqrt{t}$$

where r is the risk-free rate.$\square$

We note that these formulas do not involve the natural drift. In fact, the only "unknown" quantities are the natural volatility σ_ν and the risk-free rate r.

Example 10.2 Consider a European call option on a stock that is currently selling for \$100. The option expires in 1 year at a strike price of \$100. Suppose that the risk-free rate is 5% and that the volatility is $\sigma = 0.15$ per year. Compute the value of the call.
Solution This is simply a matter of plugging into the formula. First, we have

$$d_2 = \frac{1}{0.15}\left[0 + 0.05 - \frac{1}{2}(0.15)^2\right] \approx 0.2583$$

and

$$d_1 = d_2 + \sigma\sqrt{t} \approx 0.2583 + 0.15 = 0.4083$$

Hence, with the aid of a calculator or some other means of evaluating values of $\phi_{0,1}$, we get

$$C = 100\phi_{0,1}(0.4083) - 100e^{-0.05}\phi_{0,1}(0.2583) \approx \$8.596 \qquad \square$$

How Black–Scholes Is Used in Practice: Volatility Smiles

The assumption that the volatility σ is constant is a very unrealistic one, to say the least. This (among other things) raises questions about the quality of the

Black–Scholes option pricing formula. A great deal of research has been done to determine how the Black–Scholes formula can be used in light of the questionable assumptions about the parameters. Our intention is to very briefly discuss one method that is used in practice.

The Volatility Smile

We have said that the volatility can be estimated using historical data. However, in practice, the Black–Scholes formula is not used by simply plugging in an estimate for the volatility and grinding out option prices. Instead traders usually work with a quantity known as the *implied volatility*. Loosely speaking, this is the volatility that must be used in the Black–Scholes option pricing formula in order to make the formula reflect the actual market price at a given moment in time.

Definition *Consider a European option that has a particular market price of* M. *The* **implied volatility** *of this option is the volatility that is required in the Black–Scholes option pricing formula so that the formula gives* M. $\Box$

The implied volatility is, in effect, the market's opinion about the Black–Scholes volatility of a stock. The implied volatility is a quantity that can be computed from the Black–Scholes formula by numerical methods (that is, intelligent guessing and reguessing).

As it happens, and as further evidence that the Black–Scholes formula is not perfect, if one computes the implied volatility of otherwise identical options at various strike prices, one gets different values. But, of course, the volatility in the stock price does not depend on the strike price of an option.

Figure 10.5 shows a typical graph of implied volatility versus strike price. Because of the shape of the graph, it is known as a **volatility smile**.

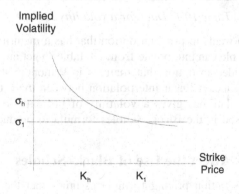

Figure 10.5: A volatility smile

Now, suppose that a historical estimate of volatility for a particular stock is σ_h. Then the Black–Scholes formula for this volatility gives an option price that matches the market price for only one strike price K_h.

For larger strike prices K_1, which correspond to out-of-the-money calls and in-the-money puts, the market price corresponds to an implied volatility σ_1 that is less than σ_h. But the Black–Scholes formula gives prices that vary *directly* with the volatility. Therefore, the larger historical volatility σ_h produces a larger option price in the Black–Scholes formula. In other words, when the Black–Scholes formula is used with a volatility based on historical data, the formula tends to *overprice* out-of-the-money calls and in-the-money puts relative to market prices.

Similarly, the Black–Scholes formula tends to *underprice* in-the-money calls and out-of-the-money puts relative to market prices, when used with a historical volatility.

The Volatility Surface

To understand how the Black–Scholes formula may be used in practice, consider Table 10.1, which shows data for a *volatility surface*; that is, a set of implied volatilities for various maturity dates and strike prices. The columns represent different strike prices expressed as a percentage of the stock price. (The data in this table is for illustration only.)

	90%	95%	100%	105%	110%
1 month	14	13	12	11.3	14.4
3 months	14.2	14.1	13.6	13.8	14.1
6 months	14.5	14.3	14.2	14.5	14.6
1 year	15.1	15	14.6	14.7	14.8
2 years	16.2	16.1	16	16.1	16.2

Table 10.1: Data for a volatility surface

Now, a trader who wants to price an option that has a maturity and strike price that is not in this table can interpolate from the table to get an implied volatility. For example, consider an option that matures in 9 months at a strike price of 95% of the stock price. A linear interpolation between the 6 month and 1 year maturity implied volatilities gives a volatility of $(14.3 + 15)/2 = 14.65$. This volatility can be used in the Black–Scholes formula to produce a price for the option in question.

How Dividends Affect the Use of Black–Scholes

The Black–Scholes option pricing formula assumes that the underlying stock does not pay a dividend. We should briefly discuss how dividends can be handled in this context.

First, we need a bit of background on dividends. Four dates are important with respect to dividends. The *declaration date* is the date that the board of directors declares a dividend. The *record date* is the date that the registrar compiles a list of current shareholders, to whom the dividend will be paid. The key point is that an investor must be on record as owning the stock on the record date or else he will not receive a dividend. The *payment date* is the date that the dividend will be paid.

Now, normal stock purchases take 3 days to clear. This is referred to as *regular way* settlement. Thus, an owner must have purchased the stock at least 3 days prior to the record date in order to be considered an *owner of record* on the record date and hence eligible for the dividend. The first date after this date is called the *ex-dividend date*. For example, if a dividend is declared as payable to stockholders of record on a given Friday, then the New York Stock Exchange (which sets the ex-dividend dates for its stocks) would declare the stock "ex-dividend" as of the opening of the market on the preceding Wednesday.

We note that when the stock goes ex-dividend, typically the stock price will decline by an amount similar to the amount of the dividend. This makes sense from the perspective that the dividends are known in advance and are therefore built into the stock's price in some way.

Now, a European option on a stock that pays a dividend can be thought of as composed of two separate processes: A risky process that represents the stock price itself and a risk-free process that represents the cash dividend payments. Thus, to price an option on such a stock, we first discount all of the forthcoming dividend payments to the present. If this amount is d, then we can think of the two components as a risky stock with initial value $S_0 - d$ and a risk-free asset with initial value d. The Black–Scholes formula can then be applied to the risky stock.

The Binomial Model from a Different Perspective: Itô's Lemma

Our approach to studying the continuous-time model was as a limiting case of the binomial model. The key was to use the martingale measure binomial models to obtain the weak limit

$$H_T \xrightarrow{\text{dist}} H_t = \left(r - \frac{\sigma_\nu^2}{2} \right) t + \sigma_\nu \sqrt{t} Z_t \tag{10.1}$$

We want to take a few minutes now to discuss what is generally considered to be a more rigorous direct approach to the continuous-time model. Since a formal discussion would require considerably more mathematical machinery than we have at hand, we will proceed informally. However, the reader may feel free to skip this discussion without loss of continuity (no pun intended).

Let us begin by taking a closer look at the notion of *rate of return* on a stock price. This term has more than one meaning. If we write

$$s = \frac{1}{T} \log\left(\frac{S_T}{S_0}\right) = \frac{1}{T} H_T$$

then

$$S_T = S_0 e^{sT}$$

and so the stock price grows at a continuously compounded rate of s. Thus, s is the *continuously compounded rate of return*.

On the other hand, the **simple rate of return** of the stock price over a short period of time $[t, t + \Delta t]$ is given by

$$\frac{\Delta S_t}{S_t} = \frac{S_{t+\Delta t} - S_t}{S_t}$$

In the limit as $\Delta t \to 0$, the simple rate of return can be thought of as a rate of return over an infinitesimal time period dt. This is more appropriately called the **instantaneous percentage return** of the stock price and is denoted by

$$\frac{dS_t}{S_t}$$

Now, the most common approach to the continuous-time model of stock prices assumes that the instantaneous percentage return is a Brownian motion process, more specifically, that

$$\frac{dS_t}{S_t} = \mu_0 \, dt + \sigma_0 \, dW_t$$

where $\{W_t \mid t \geq 0\}$ is standard Brownian motion and μ_0 and σ_0 are constants. This equation must remain somewhat vague for us, because the meanings of the differentials dS_t and dW_t are rather involved. However, we can say that the stochastic process

$$\left\{ \frac{dS_t}{S_t} \,\Big|\, t \geq 0 \right\}$$

is assumed to follow a Brownian motion process with drift μ_0 and volatility σ_0.

The previous formula can be written

$$dS_t = \mu_0 S_t dt + \sigma_0 S_t \, dW_t$$

This is an instance of what is known as a *stochastic differential equation*, whose formal solution requires some rather sophisticated mathematical machinery known as *stochastic calculus*. Nonetheless, we can wave our hands a bit to get some further insight into how this stochastic differential equation is used to develop the model.

The previous equation is a special case of the equation

$$dS_t = a(S_t, t) \, dt + b(S_t, t) \, dW_t$$

where $a(x, t)$ and $b(x, t)$ are functions of two variables. In our case

$$a(S_t, t) = \mu_0 S_t \quad \text{and} \quad b(S_t, t) = \sigma_0 S_t$$

A process $\{S_t\}$ that obeys the preceding equation is sometimes called an **Itô process**.

Now, if $f(x, t)$ is a function of two variables, then we may form the composition $f(S_t, t)$, which is a stochastic process, since $\{S_t\}$ is a stochastic process. A formula for the differential df can be obtained by applying a result from the stochastic calculus known as Itô's lemma.

Theorem 10.8 (Itô's Lemma) *Let $\{S_t\}$ follow an Itô process*

$$dS_t = a(S_t, t) \, dt + b(S_t, t) \, dW_t$$

where $\{W_t\}$ is standard Brownian motion and $a(x, t)$ and $b(x, t)$ are functions of two variables. Let $f(x, t)$ be a (sufficiently differentiable) function of two variables. Then

$$df = \left(\frac{\partial f}{\partial x} a + \frac{\partial f}{\partial t} + \frac{1}{2} \frac{\partial^2 f}{\partial x^2} b^2 \right) dt + \frac{\partial f}{\partial x} b \, dW_t \qquad \square$$

In the case at hand, we have

$$a(S_t, t) = \mu_0 S_t$$
$$b(S_t, t) = \sigma_0 S_t$$

and so Itô's lemma becomes

$$df = \left(\frac{\partial f}{\partial x} \mu_0 S_t + \frac{\partial f}{\partial t} + \frac{1}{2} \frac{\partial^2 f}{\partial x^2} \sigma_0^2 S_t^2 \right) dt + \frac{\partial f}{\partial x} \sigma_0 S_t \, dW_t$$

Let us now apply this formula to the function

$$f(x, t) = \log x$$

In this case,

$$\frac{\partial f}{\partial x}(S_t, t) = \frac{1}{S_t}, \quad \frac{\partial f}{\partial t}(S_t, t) = 0 \quad \text{and} \quad \frac{\partial^2 f}{\partial x^2}(S_t, t) = -\frac{1}{S_t^2}$$

and we get

$$d(\log S_t) = \left(\frac{1}{S_t}\mu_0 S_t - \frac{1}{2}\frac{1}{S_t^2}\sigma_0^2 S_t^2\right)dt + \frac{1}{S_t}\sigma_0 S_t\, dW_t$$
$$= \left(\mu_0 - \frac{\sigma_0^2}{2}\right)dt + \sigma_0\, dW_t$$

This says that $d(\log S_t)$ is normally distributed with mean $\left(\mu_0 - \frac{\sigma_0^2}{2}\right)dt$ and variance $\sigma_0^2\, dt$. It follows that the change in $\log S_t$ over a period $[0, t]$; that is,

$$\log S_t - \log S_0$$

being the sum of independent normal random variables, is also normally distributed with mean

$$\sum\left(\mu_0 - \frac{\sigma_0^2}{2}\right)dt = \left(\mu_0 - \frac{\sigma_0^2}{2}\right)t$$

and variance

$$\sum \sigma_0^2 dt = \sigma_0^2 t$$

Hence,

$$H_t = \log\left(\frac{S_t}{S_0}\right) = \log S_t - \log S_0$$
$$= \left(\mu_0 - \frac{\sigma_0^2}{2}\right)t + \sigma_0\sqrt{t}Z_t$$

where Z_t is standard normal. Thus, we see that the stochastic differential equation leads to the same weak limit (10.1).

Exercises

1. A stock has the initial price of $50. Over a five-day period, the stock price at day's end is given by $49.82, $50.02, $49.69, $49.34, $50.10. Estimate the natural drift and volatility.
2. Consider a European call option on a stock that is currently selling for $80. The option expires in 1 year at a strike price of $80. Suppose that the risk-free rate is 5% and that the volatility is $\sigma = 0.1$ per year. Compute the value of the call.
3. Consider a European put option on a stock that is currently selling for $50. The option expires in 1 year at a strike price of $51. Suppose that the risk-free rate is 4% and that the volatility is $\sigma = 0.15$ per year. Compute the value of the put.
4. Prove that

$$\mathrm{Var}_p(X_{p,T,i}) = 1$$

5. Prove that

$$\mathcal{E}_p(Q_s) = \sigma_s \frac{t}{\sqrt{\Delta t}} \frac{p-s}{\sqrt{s(1-s)}}$$

$$\mathrm{Var}_p(Q_s) = \sigma_s^2 t \frac{p(1-p)}{s(1-s)}$$

6. Let X be a random variable with a lognormal distribution; that is, $Y = \log X$ is normally distributed with mean a and variance b^2. Show that
   a) $\mathcal{E}(X) = e^{a+\frac{1}{2}b^2}$
   b) $\mathrm{Var}(X) = e^{2a+b^2}(e^{b^2}-1)$
   Apply this to the random variable $X = S_t/S_0$ to deduce that
   c) $\mathcal{E}_\pi(S_t) = S_0 e^{rt}$
   d) $\mathrm{Var}_\pi(S_t) = (S_0 e^{rt})^2(e^{\sigma_\nu^2 t}-1)$
7. Show that the function $f(x) = \max\{K - S_0 e^x, 0\}$ is continuous and bounded on $\mathbb{R}$.
8. Show that $\phi_{0,1}(-t) = 1 - \phi_{0,1}(t)$ for the standard normal distribution function. *Hint*: Draw a picture using the graph of the standard normal density function.
9. Show using l'Hopital's rule that

$$\lim_{T\to\infty} \pi_T = \lim_{\Delta t\to 0} \left(\frac{e^{(r-\mu)\Delta t + \frac{p}{\sqrt{pq}}\sigma\sqrt{\Delta t}} - 1}{e^{\frac{1}{\sqrt{pq}}\sigma\sqrt{\Delta t}} - 1} \right) = p$$

Hint: Write $x = (\Delta t)^2$.

10. Suppose we assume that there is a martingale probability measure in the limiting case when $T \to \infty$ and that the Fundamental Theorem of Asset Pricing holds in this limiting case. If we denote this martingale measure by Π then

$$S_0 = e^{-rt}\mathcal{E}_\Pi(S_t) = e^{-rt}\mathcal{E}_\Pi(S_0 e^{\mu_\nu t + \sigma_\nu \sqrt{t}Z_t})$$

Evaluate the last expression to show that

$$\mu_\nu = r - \frac{1}{2}\sigma_\nu$$

11. Let N_U be the random variable representing the *number of* up-ticks in stock price over the lifetime of the binomial model. Show that

$$S_{t,T} = S_0 e^{N_U(\log u - \log d) + T\log d}$$

and so

$$H_{t,T} = N_U(\log u - \log d) + T\log d$$

Show that

$$\mathcal{E}(H_{t,T}) = T\nu(\log u - \log d) + T\log d$$
$$\text{Var}(H_{t,T}) = T\nu(1 - \nu)(\log u - \log d)^2$$

12. Standardize the random variable $H_{t,T}$ to show that

$$H^*_{t,T} = N^*_U$$

Hence, the random variables $H^*_{t,T}$ are standardized binomial random variables.

13. Using the Black–Scholes formulas, show that the values of a put and a call increase as the volatility σ increases. Looking at the profit curves for a long put and long call, explain why this makes sense. Does the same effect obtain for the owner of a stock?

14. Show that the probability that a European call with strike price K and expiration date t will expire in or at the money is

$$\phi_{0,1}\left(\frac{\log(S_T/K) - t(r - \sigma_\nu^2/2)}{\sigma\sqrt{t}} \right)$$

where $\phi_{0,1}$ is the standard normal distribution.

Appendix A
Convexity and the Separation Theorem

In this appendix, we develop the material on convexity required for the proof of the First Fundamental Theorem of Asset Pricing. Let us begin with some basic facts, notation and terminology.

We use the notation $\langle \alpha, \beta \rangle$ for the standard inner product in $\mathbb{R}^n$. The **Cauchy–Schwarz inequality** states that

$$|\langle \alpha, \beta \rangle| \leq \|\alpha\|\|\beta\|$$

with equality if and only if $\beta = a\alpha$ for some $a \in \mathbb{R}$. Note that this inequality shows that for any $\beta \in \mathbb{R}^n$, the inner product function

$$f_\beta(\alpha) = \langle \alpha, \beta \rangle$$

is continuous, since $\alpha_n \to \alpha$ implies that

$$|f_\beta(\alpha_n) - f_\beta(\alpha)| = |f_\beta(\alpha_n - \alpha)| = |\langle \alpha_n - \alpha, \beta \rangle| \leq \|\alpha_n - \alpha\|\|\beta\| \to 0$$

It follows from the Riesz Representation Theorem (Theorem 7.3) that any linear functional on $\mathbb{R}^n$ is continuous. It also follows that the norm function

$$f(\alpha) = \|\alpha\|$$

is continuous on $\mathbb{R}^n$.

If $\alpha \in \mathbb{R}^n$ and $S \subseteq \mathbb{R}^n$, we write

$$\langle \alpha, S \rangle = \{\langle \alpha, \sigma \rangle \mid \sigma \in S\}$$

and write

$$\langle \alpha, S \rangle \leq b$$

to mean that $\langle \alpha, \sigma \rangle \leq b$ for all $\sigma \in S$.

An **affine subspace** of $\mathbb{R}^n$ is a set of the form $\beta + S$ where S is a subspace of $\mathbb{R}^n$. The affine subspace $\beta + S$ is **nontrivial** if $S \neq \{0\}$.

If $C \subseteq \mathbb{R}^n$, we say that $\alpha \in C$ is a **smallest** member of C if the distance $\|\alpha\|$ from α to the origin satisfies

$$\|\alpha\| \leq \|x\|$$

for all $x \in C$. Note that a set can have more than one smallest element.

The set

$$\mathbb{R}^n_+ = \{(x_1, \ldots, x_n) \mid x_i \geq 0\}$$

is called the **nonnegative orthant** in $\mathbb{R}^n$ and the set

$$\mathbb{R}^n_{++} = \{(x_1, \ldots, x_n) \mid x_i > 0\}$$

is called the **positive orthant** in $\mathbb{R}^n$.

Convex, Closed and Compact Sets

We shall need the following concepts.

Definition *Let $X \subseteq \mathbb{R}^n$.*
1) *Let $x_1, \ldots, x_k \in \mathbb{R}^n$. A linear combination of the form*

$$t_1 x_1 + \cdots + t_k x_k$$

where

$$t_1 + \cdots + t_k = 1 \quad and \quad 0 \leq t_i \leq 1$$

is called a **convex combination** *of the vectors $x_1, \ldots, x_k$.*
2) *X is* **convex** *if whenever $x, y \in X$, the line segment between x and y also lies in X, in symbols,*

$$x, y \in X \quad \Rightarrow \quad \{sx + ty \mid s + t = 1, 0 \leq s, t \leq 1\} \subseteq X$$

3) *X is a* **cone** *if X is nonempty and*

$$x \in X, \quad \alpha > 0 \quad \Rightarrow \quad \alpha x \in X$$

4) *X is* **closed** *if whenever (x_n) is a sequence in X that converges in $\mathbb{R}^n$, then the limit is in X.*
5) *X is* **compact** *if it is both closed and bounded.* $\square$

Example A.1 We leave it as an exercise to show that the nonnegative orthant $\mathbb{R}^n_+$ is a closed, convex cone.$\square$

We will also have need of the following facts from analysis.

1) A continuous function that is defined on a compact set X in $\mathbb{R}^n$ takes on its maximum and minimum values at some points within the set X.

2) A subset X of $\mathbb{R}^n$ is compact if and only if every sequence in X has a subsequence that converges in X.

Theorem A.1 *Let X and Y be subsets of $\mathbb{R}^n$ and let*

$$X + Y = \{a + b \mid a \in X, b \in Y\}$$

1) If X and Y are convex, then so is $X + Y$.
2) If X is compact and Y is closed, then $X + Y$ is closed. However, the sum of two closed sets need not be closed.
3) Any subspace S of $\mathbb{R}^n$ is closed and convex.
4) As to smallest points, a nonempty convex subset C of $\mathbb{R}^n$ has at most one smallest point and a nonempty closed convex subset C of $\mathbb{R}^n$ has exactly one smallest point.
Proof. We leave proof of part 1) to the reader. For part 2), a sequence in $X + Y$ has the form $(a_n + b_n)$, where $a_n \in X$ and $b_n \in Y$ for all $n = 0, 1, 2, \ldots$. We must show that if $(a_n + b_n) \to c$, then $c \in X + Y$. Since X is compact, the sequence (a_n) has a convergent subsequence (a_{n_k}) whose limit α lies in X. Since $a_{n_k} + b_{n_k} \to c$ and $a_{n_k} \to \alpha$, we can conclude that $b_{n_k} \to c - \alpha$. But Y is closed and so $c - \alpha \in Y$, whence $c = \alpha + (c - \alpha) \in X + Y$, as desired.

For the second statement in part 2), let

$$X = \{n + 1/n \mid n \geq 1\} \quad \text{and} \quad Y = \{-n \mid n \geq 1\}$$

which are both closed. Then $\{1/n\} \subseteq X + Y$ but $0 \notin X + Y$ and so $X + Y$ is not closed.

For part 3), it is clear that a subspace of $\mathbb{R}^n$ must be convex. To show that it is closed, extend an orthonormal basis $\{u_1, \ldots, u_k\}$ for S to an orthonormal basis $\{u_1, \ldots, u_k, v_1, \ldots, v_m\}$ for $\mathbb{R}^n$. If $x_n \in S$, then

$$x_n = \sum_{i=1}^{k} a_{n,i} u_i$$

Suppose that $(x_n) \to y$, where

$$y = \sum_{i=1}^{k} c_i u_i + \sum_{i=1}^{m} d_i v_i$$

Then

$$|x_n - y|^2 = \sum_{i=1}^{k} |a_{n,i} - c_i|^2 + \sum_{i=1}^{m} |d_i|^2$$

and so $|d_i|^2 \leq |x_n - y|^2$ for all i and since $(x_n) \to y$, it follows that $d_i = 0$ for all i. Hence, $y \in S$.

For part 4), if $0 \in C$, then it is clearly the unique smallest point of C, so assume that $0 \notin C$. For purposes of contradiction, assume that α and β are distinct smallest points of C and let $\gamma = (\alpha + \beta)/2 \in C$. Since α is not a scalar multiple of β, the Cauchy–Schwarz inequality gives

$$
\begin{aligned}
4\|\gamma\|^2 &= \|(\alpha + \beta)\|^2 \\
&= \|\alpha\|^2 + 2\langle \alpha, \beta \rangle + \|\beta\|^2 \\
&= 2\|\alpha\|^2 + 2\langle \alpha, \beta \rangle \\
&< 2\|\alpha\|^2 + 2\|\alpha\|\|\beta\| \\
&= 4\|\alpha\|^2
\end{aligned}
$$

and so $\|\gamma\| < \|\alpha\|$, which contradicts the choice of α as a smallest point of C.

For the second part, assume that C is closed and the norm function

$$
d(x) = \|x\|
$$

on C is continuous. Although C need not be compact, if we choose a real number r large enough so that the closed ball $B_r(0) = \{x \in \mathbb{R}^n \mid \|x\| \le r\}$ of radius r about the origin intersects C nontrivially, that is,

$$
C \cap B_r(0) \neq \emptyset
$$

then $C \cap B_r(0)$ is nonempty, closed and bounded and so is compact. The distance function d therefore achieves its minimum at a point $\alpha \in C \cap B_r(0)$, which is also a smallest point of C. $\square$

Convex Hulls

We will have use for the notion of a convex hull.

Definition *The **convex hull** of a set $S = \{x_1, \ldots, x_k\}$ of vectors in $\mathbb{R}^n$ is the smallest convex set in $\mathbb{R}^n$ that contains the vectors $x_1, \ldots, x_k$. We denote the convex hull of S by $\mathcal{C}(S)$.* $\square$

The convex hull is easily characterized as follows.

Theorem A.2 *If $S = \{x_1, \ldots, x_k\} \subseteq \mathbb{R}^n$, then the convex hull $\mathcal{C}(S)$ is the set*

$$
\Delta = \left\{ t_1 x_1 + \cdots + t_k x_k \,\middle|\, 0 \le t_i \le 1, \sum t_i = 1 \right\}
$$

of all convex combinations of vectors in S.
Proof. We leave it to the reader to show that Δ is convex. Since $x_i \in \Delta$ for all i, the definition of $\mathcal{C}(S)$ implies that $\mathcal{C}(S) \subseteq \Delta$. For the reverse inclusion, we must show that if D is convex and $S \subseteq D$, then $\Delta \subseteq D$. Note that D contains all convex combinations of any two vectors in S. For a convex combination $t_1 x_1 + t_2 x_2 + t_3 x_3$ of three vectors in S, we can write

$$x = t_1 x_1 + t_2 x_2 + t_3 x_3 = t_1 x_1 + (t_2 + t_3)\left(\frac{t_2}{t_2 + t_3}x_2 + \frac{t_3}{t_2 + t_3}x_3\right)$$
$$= t_1 x_1 + (t_2 + t_3)y$$

where y is defined as the expression in parentheses above. But y is a convex combination of x_2 and x_3 and so $y \in D$. Then x is a convex combination of x_1 and y and so $x \in D$. Thus, any convex combination of three vectors in S is in D. An inductive argument along these lines, which we leave as an exercise, can be used to furnish a complete proof that any convex combination of vectors in S is in D; that is, $\Delta \subseteq D$. $\square$

Linear and Affine Hyperplanes

A **linear hyperplane** in $\mathbb{R}^n$ is an $(n-1)$-dimensional subspace of $\mathbb{R}^n$. As such, it is the solution set of a linear equation of the form

$$a_1 x_1 + \cdots + a_n x_n = 0$$

or

$$\langle \alpha, x \rangle = 0$$

where $\alpha = (a_1, \ldots, a_n)$ is nonzero and $x = (x_1, \ldots, x_n)$. Geometrically speaking, this is the set of all vectors in $\mathbb{R}^n$ that are orthogonal to the nonzero vector α.

An **(affine) hyperplane** is a linear hyperplane that has been translated by a vector $\beta = (b_1, \ldots, b_n)$. Thus, it is the solution set to an equation of the form

$$\langle \alpha, x - \beta \rangle = 0$$

or equivalently,

$$\langle \alpha, x \rangle = \langle \alpha, \beta \rangle$$

Letting $b = \langle \alpha, \beta \rangle$, we can write a hyperplane in the form

$$\mathcal{H}(\alpha, b) = \{x \in \mathbb{R}^n \mid \langle \alpha, x \rangle = b\}$$

where $\alpha \in \mathbb{R}^n$ and $b \in \mathbb{R}$.

Also, it is clear that, by dividing both sides by $\|\alpha\|$ and multiplying both sides by -1 if necessary, all hyperplanes have the form $\mathcal{H}(\alpha, b)$ where $\|\alpha\| = 1$ and $b \geq 0$. In this case, $b\alpha \in \mathcal{H}(\alpha, b)$ and so for any $x \in \mathcal{H}(\alpha, b)$, we have

$$\langle x - b\alpha, b\alpha \rangle = \langle x, b\alpha \rangle - \langle b\alpha, b\alpha \rangle = b^2 - b^2 = 0$$

Hence, the Pythagorean Theorem gives

$$\|x - b\alpha\|^2 + \|b\alpha\|^2 = \|x\|^2$$

which implies that $b\alpha$ is the *unique* element of $\mathcal{H}(\alpha, b)$ of smallest norm.

We can now catalog all hyperplanes as follows. The linear hyperplanes are uniquely cataloged by $\mathcal{H}(\alpha, 0)$ as α ranges over all *unit vectors*. Hyperplanes that do not pass through the origin are uniquely cataloged by their unique smallest member α as

$$\mathcal{H}\left(\frac{\alpha}{\|\alpha\|}, \|\alpha\|\right) = \mathcal{H}(\alpha, \|\alpha\|^2) = \{x \in \mathbb{R}^n \mid \langle x, \alpha \rangle = \|\alpha\|^2\}$$

A hyperplane divides $\mathbb{R}^n$ into two **closed half-spaces**

$$\mathcal{H}_{\geq}(\alpha, b) = \{x \in \mathbb{R}^n \mid \langle \alpha, x \rangle \geq b\}$$
$$\mathcal{H}_{\leq}(\alpha, b) = \{x \in \mathbb{R}^n \mid \langle \alpha, x \rangle \leq b\}$$

where $\mathcal{H}_{\leq}(\alpha, b)$ contains the origin. The corresponding **open half-spaces** are

$$\mathcal{H}_{>}(\alpha, b) = \{x \in \mathbb{R}^n \mid \langle \alpha, x \rangle > b\}$$
$$\mathcal{H}_{<}(\alpha, b) = \{x \in \mathbb{R}^n \mid \langle \alpha, x \rangle < b\}$$

It seems intuitively clear that if a nontrivial affine subspace $\beta + S$ is contained within a half-space $\mathcal{H}_{>}(\alpha, b)$, then S must be parallel to the bounding plane $\mathcal{H}(\alpha, b)$ of the half-space. To prove this, the containment $\beta + S \subseteq \mathcal{H}_{>}(\alpha, b)$ is equivalent to the condition

$$\langle \alpha, \beta + S \rangle > b$$

and so for all nonzero $s \in S$ and all real numbers a,

$$\langle \alpha, \beta \rangle + a\langle \alpha, s \rangle = \langle \alpha, \beta + as \rangle > b$$

But this can happen only if $\langle \alpha, s \rangle = 0$ and so

$$\alpha \perp S$$

A similar argument holds for the other open half-space $\mathcal{H}_{<}(\alpha, b)$ and for the two closed half-spaces.

Theorem A.3 *If a nontrivial affine subspace $\beta + S$ of $\mathbb{R}^n$ lies in either open or closed half-space determined by a hyperplane $\mathcal{H}(\alpha, b)$, then*

$$\alpha \perp S \qquad \qquad \square$$

Separation

Now that we have the preliminaries out of the way, we can turn to the main issue of separation.

Definition *Two nonempty subsets X and Y of $\mathbb{R}^n$ are* **completely separated** *by a hyperplane $\mathcal{H}(\alpha, b)$ if X lies in one open half-space determined by $\mathcal{H}(\alpha, b)$ and Y lies in the other; that is, if*

$$\langle \alpha, X \rangle < b < \langle \alpha, Y \rangle \quad or \quad \langle \alpha, Y \rangle < b < \langle \alpha, X \rangle \qquad \square$$

The following simple result is key to separation theorems.

Theorem A.4 *As shown in Figure A.1, if $\beta \in \mathbb{R}^n$ lies in the open half-space $\mathcal{H}_<(\alpha, \|\alpha\|^2)$ containing the origin; that is, if*

$$\langle \alpha, \beta \rangle < \|\alpha\|^2$$

then there is a point on the open line segment from α to β that is closer to the origin than α.
Proof. For any $t \in \mathbb{R}$,

$$
\begin{aligned}
\|(1-t)\alpha &+ t\beta\|^2 \\
&= (1-t)^2\|\alpha\|^2 + 2t(1-t)\langle \beta, \alpha \rangle + t^2\|\beta\|^2 \\
&= (\|\beta\|^2 + \|\alpha\|^2 - 2\langle \beta, \alpha \rangle)t^2 + 2(\langle \beta, \alpha \rangle - \|\alpha\|^2)t + \|\alpha\|^2
\end{aligned}
$$

This is less than $\|\alpha\|^2$ if and only if

$$(\|\beta\|^2 + \|\alpha\|^2 - 2\langle \beta, \alpha \rangle)t^2 + 2(\langle \beta, \alpha \rangle - \|\alpha\|^2)t < 0$$

But $\langle \beta, \alpha \rangle - \|\alpha\|^2 < 0$ and so if t is a small enough positive number, then this inequality does hold. $\square$

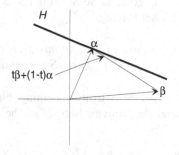

Figure A.1

We can now state our first separation theorem, which is the basis for other separation theorems.

Theorem A.5
1) *Let C be a convex subset of $\mathbb{R}^n$ that has a nonzero unique smallest point $\alpha \in C$. Then C lies in the closed half-space $\mathcal{H}_\geq(\alpha, \|\alpha\|^2)$ and so C and 0 are completely separated by the hyperplane $\mathcal{H}(\alpha/2, \|\alpha/2\|^2)$.*

2) *In particular, part 1) applies to nonempty closed, convex subsets C of $\mathbb{R}^n$
    that do not contain the origin.*

Proof. Part 1) follows immediately from Theorem A.4, since if $\beta \in C$ lies in the
open half-space $\mathcal{H}_<(\alpha, \|\alpha\|^2)$, then there is a point on the line segment from β to
α that is closer to the origin then α. However, the convexity of C and the
minimality of $\|\alpha\|$ implies that this cannot happen and so $C \subseteq \mathcal{H}_{\geq}(\alpha, \|\alpha\|^2)$.
Part 2) follows from Theorem A.1.$\square$

We next consider separating compact convex subsets from subspaces.

Theorem A.6 *Let C be a compact convex subset of $\mathbb{R}^n$ and let S be a subspace
of $\mathbb{R}^n$ such that $C \cap S = \emptyset$. Then there is a nonzero $\alpha \in \mathbb{R}^n$ for which*

$$\alpha \perp S \quad and \quad \langle \alpha, C \rangle > \|\alpha\|^2$$

Hence, the hyperplane $\mathcal{H}(\alpha, \|\alpha\|^2)$ completely separates S and C.
Proof. The set

$$A = S + C$$

is closed and convex and $0 \notin A$ since $C \cap S = \emptyset$. Hence, Theorem A.5 implies
that there exists a hyperplane $\mathcal{H}(\alpha, \|\alpha\|^2)$ for $\alpha \neq 0$ that completely separates A
from 0; that is, $A \subseteq \mathcal{H}_>(\alpha, \|\alpha\|^2)$. Since $C \subseteq A$, it follows that
$C \subseteq \mathcal{H}_>(\alpha, \|\alpha\|^2)$,   whence $\langle \alpha, C \rangle > \|\alpha\|^2$.   Furthermore,   since
$c + S \subseteq \mathcal{H}_>(\alpha, \|\alpha\|^2)$ for any $c \in C$, Theorem A.3 implies that $S \perp \alpha.\square$

Now we come to our main goal, at least with respect to proving the First
Fundamental Theorem of Asset Pricing.

Theorem A.7 *Let S be a subspace of $\mathbb{R}^n$ that intersects the nonnegative orthant
$\mathbb{R}_+^n$ trivially; that is, $S \cap \mathbb{R}_+^n = \{0\}$. Then $S^\perp$ contains a strongly positive
vector.*
Proof. We would like to separate S from something, but we cannot separate it
from $\mathbb{R}_+^n$. Consider instead the convex hull Δ of the standard basis vectors
$\epsilon_1, \ldots, \epsilon_n$ in $\mathbb{R}_+^n$

$$\Delta = \{t_1\epsilon_1 + \cdots + t_n\epsilon_n \mid 0 \leq t_i \leq 1, \Sigma t_i = 1\}$$

(In $\mathbb{R}^2$, this is the closed line segment from $(0, 1)$ to $(1, 0)$ and in $\mathbb{R}^3$, this is the
triangle with vertices $(1, 0, 0)$, $(0, 1, 0)$ and $(0, 0, 1)$.) It is clear that
$\Delta \subseteq \mathbb{R}_+^n \setminus \{0\}$ and so $\Delta \cap S = \emptyset$. Also, Δ is convex and closed and bounded
and therefore compact. Hence, Theorem A.6 implies that there is a nonzero
vector $\alpha = (a_1, \ldots, a_n)$ such that

$$\alpha \perp S \quad and \quad \langle \alpha, \Delta \rangle > \|\alpha\|^2$$

In particular, since $\epsilon_i \in \Delta$ for all i, we have

$$a_i = \langle \alpha, \epsilon_i \rangle > \|\alpha\|^2 > 0$$

and so α is strongly positive, as desired. $\square$

The following theorem is an analog to Theorem A.7 for strictly positive vectors.

Theorem A.8 *Let S be a subspace of $\mathbb{R}^n$ for which $S \cap \mathbb{R}^n_{++} = \emptyset$. Then $S^\perp$ contains a strictly positive vector.*
Proof. First note that $S^\perp$ contains a strictly positive vector α if and only if it contains a strictly positive vector whose coordinates sum to 1. (Just divide α by the sum of its coordinates.)

Let $\mathcal{B} = \{B_1, \dots, B_k\}$ be a basis for S and consider the matrix

$$M = (B_1 \mid B_2 \mid \cdots \mid B_k)$$

whose columns are the basis vectors in $\mathcal{B}$. Let the rows of M be denoted by $R_1, \dots, R_m$. Note that $R_i \in \mathbb{R}^k$, where $k = \dim(S)$.

Now, $\alpha = (a_1, \dots, a_m) \in S^\perp$ if and only if $\langle \alpha, B_i \rangle = 0$ for all i, which is equivalent to the matrix equation

$$\alpha M = 0$$

or the vector equation

$$a_1 R_1 + \cdots + a_m R_m = 0$$

Hence, $S^\perp$ contains a strictly positive vector $\alpha = (a_1, \dots, a_m)$ whose coordinates sum to 1 if and only if this equation holds for coefficients $a_i \geq 0$ satisfying $\Sigma a_i = 1$. In other words, $S^\perp$ contains a strictly positive vector if and only if 0 is contained in the convex hull C of the vectors $R_1, \dots, R_m$ in $\mathbb{R}^k$.

Now, for the purposes of contradiction, let us assume that $0 \notin C$. Since C is closed and convex, Theorem A.5 implies that there is a nonzero vector $\beta = (b_1, \dots, b_k) \in \mathbb{R}^k$ for which

$$\langle \beta, C \rangle > \|\beta\|^2 > 0$$

Now, consider the vector

$$v = b_1 B_1 + \cdots + b_k B_k \in S$$

If we denote the jth coordinate of R_i by $R_{i,j}$, then the ith coordinate of v is

$$\begin{aligned}
\langle v, \epsilon_i \rangle &= b_1 \langle B_1, \epsilon_i \rangle + \cdots + b_k \langle B_k, \epsilon_i \rangle \\
&= b_1 R_{i,1} + \cdots + b_k R_{i,k} \\
&= \langle \beta, R_i \rangle > 0
\end{aligned}$$

and so $v \in S \cap \mathbb{R}^n_{++}$, which is a contradiction. Hence, $0 \in C$ and so $S^{\perp}$ contains a strictly positive vector. $\square$

Appendix B
Closed, Convex Cones

In this appendix, we use the notation of Chapter 7.

Our goal in this appendix is to derive the mathematics necessary to prove that for any vector $W \in \mathbb{R}^m$, the minimum dominating price of W is equal to the maximum extension price of W,

$$P = \min\{\mathcal{I}(X) \mid X \in \mathcal{D}_W\} = \max\{f(W) \mid f \in \mathcal{E}_{\geq 0}(\mathcal{I})\}$$

where $\mathcal{I}$ is the initial pricing functional, $\mathcal{D}_W$ is the set of all attainable vectors that dominate W and $\mathcal{E}_{\geq 0}(\mathcal{I})$ is the set of all nonnegative extensions of $\mathcal{I}$. We will also prove that there is an extension $f \in \mathcal{E}_{>0}(W)$ and a dominating vector $Y \in \mathcal{D}_W$ for which

$$P = \mathcal{I}(Y) = f(W)$$

We begin by proving that

$$-\infty < \max\{f(W) \mid f \in \mathcal{E}_{\geq 0}(\mathcal{I})\} \leq \min\{\mathcal{I}(X) \mid X \in \mathcal{D}_W\} < \infty \quad \text{(B.1)}$$

Indeed, since $\mathcal{E}_{\geq 0}(\mathcal{I})$ and $\mathcal{D}_W$ are nonempty, the first and last inequalities hold. As to the middle inequality, if $X \in \mathcal{D}_W$ and $f \in \mathcal{E}_{\geq 0}(\mathcal{I})$, then $W \leq X$ implies that

$$f(W) \leq f(X) = \mathcal{I}(X)$$

Taking the maximum over all nonnegative extensions gives

$$\max\{f(W) \mid f \in \mathcal{E}_{\geq 0}(\mathcal{I})\} \leq \mathcal{I}(X)$$

and taking the minimum over all elements $X \in \mathcal{D}_W$ gives

$$\max\{f(W) \mid f \in \mathcal{E}_{\geq 0}(\mathcal{I})\} \leq \min\{\mathcal{I}(X) \mid X \in \mathcal{D}_W\}$$

In order to prove the reverse inequality, we need some additional background.

Closed, Convex Cones

Closed, convex cones in $\mathbb{R}^n$ play a key role in our discussion. A closed, convex cone $\mathcal{K}$ in $\mathbb{R}^2$ has the form shown in Figure B.1.

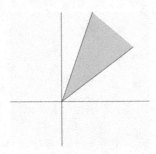

Figure B.1

It is not hard to see that $\mathcal{K}$ is equal to the intersection of all closed half-spaces of the form $\mathcal{H}_\geq(\alpha, 0)$ that contain $\mathcal{K}$. (Half-spaces are discussed in Appendix A.) Put another way, a vector $X \in \mathbb{R}^2$ is in K if and only if X is in every closed half-space $\mathcal{H}_\geq(\alpha, 0)$ containing K. This is also true in higher dimensions. We remind the reader that the inequality $\langle \alpha, \mathcal{K} \rangle \geq 0$ means that $\langle \alpha, K \rangle \geq 0$ for all $K \in \mathcal{K}$.

Theorem B.1 *If $\mathcal{K}$ is a closed, convex cone in $\mathbb{R}^n$, then the following are equivalent for any $X \in \mathbb{R}^n$:*

1) $X \in \mathcal{K}$
2) For all $\alpha \in \mathbb{R}^n$,

$$\langle \alpha, \mathcal{K} \rangle \geq 0 \quad \Rightarrow \quad \langle \alpha, X \rangle \geq 0$$

Thus, $\mathcal{K}$ is the intersection of all half-spaces containing $\mathcal{K}$.
Proof. To see that 1) implies 2), if $X \in \mathcal{K}$, then $\langle \alpha, \mathcal{K} \rangle \geq 0$ certainly implies $\langle \alpha, X \rangle \geq 0$. For the converse, suppose that 1) does not hold; that is, suppose that $X \notin K$. Then the set

$$\mathcal{K} - X = \{K - X \mid K \in \mathcal{K}\}$$

is closed and convex but does not contain the origin and so Theorem A.5 implies that there is an $\alpha \in \mathbb{R}^n$ for which

$$\langle \alpha, \mathcal{K} - X \rangle > 0$$

that is,

$$\langle \alpha, \mathcal{K} \rangle > \langle \alpha, X \rangle$$

Note that $0 \in \mathcal{K}$ implies that $\langle \alpha, X \rangle < 0$. Then it cannot be the case that $\langle \alpha, K \rangle < 0$ for any $K \in \mathcal{K}$, since $K \in \mathcal{K}$ implies that $aK \in \mathcal{K}$ for all $a \geq 0$ and so

$$0 > a\langle \alpha, K \rangle > \langle \alpha, X \rangle$$

which is false if a is sufficiently large. Thus, $\langle \alpha, K \rangle \geq 0$ and $\langle \alpha, X \rangle < 0$, which says that 2) does not hold. Thus, 2) implies 1). $\square$

We leave it to the reader to prove that the nonnegative orthant $\mathbb{R}^m_+$ is a closed, convex cone. In fact, so is any linear image of $\mathbb{R}^m_+$. In particular, if A is an $n \times m$ real matrix, then multiplication by A is a linear operator from $\mathbb{R}^m$ to $\mathbb{R}^n$ and its image $A(\mathbb{R}^m_+)$ is a closed, convex cone in $\mathbb{R}^n$.

Theorem B.2 *If A is an $n \times m$ real matrix, then $A(\mathbb{R}^m_+)$ is a closed, convex cone in $\mathbb{R}^n$.*
Proof. The theorem is certainly true for the zero matrix, so we may assume that $A \neq 0$. If the columns of A are

$$\mathcal{A} = \{A_1, \ldots, A_m\}$$

then $A(\mathbb{R}^m_+)$ is the set

$$K(\mathcal{A}) = \left\{ \sum_{i=1}^{m} x_i A_i \,\middle|\, x_i \geq 0 \right\}$$

We leave it to the reader to show that $K(\mathcal{A})$ is a convex cone. To see that $K(\mathcal{A})$ is closed, suppose that $X_k \in K(\mathcal{A})$ and that $X_k \to Y$. We must show that $Y \in K(\mathcal{A})$.

Suppose first that the vectors in $\mathcal{A}$ are linearly independent and extend $\mathcal{A}$ to a basis for $\mathbb{R}^n$, say

$$\mathcal{B} = \{A_1, \ldots, A_m, A_{m+1}, \ldots, A_n\}$$

Hence, we may write

$$X_k = x_{k,1} A_1 + \cdots + x_{k,n} A_n$$

where $x_{k,i} \geq 0$ and $x_{k,i} = 0$ for $i > m$. Also,

$$Y = y_1 A_1 + \cdots + y_n A_n$$

for $y_i \in \mathbb{R}$. Let $\{\delta_1, \ldots, \delta_n\}$ be the dual basis for $\mathcal{B}$; that is, $\delta_k : \mathbb{R}^n \to \mathbb{R}$ is the linear functional defined by

$$\delta_k(A_j) = \begin{cases} 1 & \text{if } j = k \\ 0 & \text{otherwise} \end{cases}$$

Since each δ_i is continuous (this issue is discussed in Appendix A), we have

$$\delta_i(X_k) \to \delta_i(Y)$$

for all $1 \leq i \leq n$; that is,

$$x_{k,i} \to y_i$$

But $x_{k,i} \geq 0$ for all k and i and so $y_i \geq 0$ for all i. Also, $x_{k,i} = 0$ for all k and all $i > m$ and so $y_i = 0$ for $i > m$. Hence, $Y \in K(\mathcal{A})$.

Next, we consider the case where the vectors in $\mathcal{A}$ are not linearly independent. We will show that for each $X \in K(\mathcal{A})$, there is a linearly independent subset $\mathcal{A}_X$ of $\mathcal{A}$ for which $X \in K(\mathcal{A}_X)$.

Suppose for a moment that this has been proved. Then $X_k \in K(\mathcal{A}_{X_k})$, where $\mathcal{A}_{X_k}$ is a linearly independent subset of $\mathcal{A}$. But there are only a finite number of linearly independent subsets of $\mathcal{A}$ and so there is some linearly independent subset $\mathcal{A}'$ of $\mathcal{A}$ for which there is an infinite subsequence $\{X_{k_i}\}$ of $\{X_k\}$ satisfying $X_{k_i} \in K(\mathcal{A}')$ for all k_i. Since $X_{k_i} \to Y$, the first part of this proof shows that $Y \in K(\mathcal{A}') \subseteq K(\mathcal{A})$, as desired.

Thus, we are left with proving that for each $X \in K(\mathcal{A})$, there is a linearly independent subset $\mathcal{A}_X$ of $\mathcal{A}$ for which $X \in K(\mathcal{A}_X)$. Among all nonempty subsets $\mathcal{B} \subseteq \mathcal{A}$, choose a smallest subset $\mathcal{B}$ for which $X \in K(\mathcal{B})$. By reindexing if necessary, we may assume that $\mathcal{B} = \{A_1, \ldots, A_p\}$ with $p \leq m$ and that

$$X = x_1 A_1 + \cdots + x_p A_p$$

where $x_i > 0$ for all i. We will show that $\mathcal{B}$ is linearly independent. Suppose to the contrary that there exist coefficients a_i, not all 0, for which

$$a_1 A_1 + \cdots + a_p A_p = 0$$

We may also assume that at least one a_i is positive, say $a_u > 0$. Writing

$$A_u = -\frac{1}{a_u} \sum_{k \neq u} a_k A_k$$

and substituting gives

$$X = \sum_{k=1}^{p} x_k A_k = \sum_{k \neq u} \left(x_k - \frac{x_u}{a_u} a_k \right) A_k$$

We want to find an $a_u > 0$ for which each of these coefficients is nonnegative since then

$$X \in K(\mathcal{B} \setminus \{A_u\})$$

and this contradicts the minimality of $\mathcal{B}$, thus proving that $\mathcal{B}$ is linearly independent.

Now, if $a_k = 0$, then the coefficient is $x_k \geq 0$. On the other hand, if $a_k \neq 0$, then we can write the kth coefficient in the form

$$c_k = a_k \left(\frac{x_k}{a_k} - \frac{x_u}{a_u} \right)$$

If $a_k < 0$, then $x_k / a_k \leq 0$ and $x_u / a_u \leq 0$ and so $c_k \geq 0$. For those $a_k > 0$, we need

$$\frac{x_k}{a_k} \geq \frac{x_u}{a_u}$$

and so this guides our choice of u. In particular, among all quotients x_k / a_k with $a_k > 0$, simply choose u so that x_u / a_u is smallest. Thus, $\mathcal{B}$ is linearly independent and the proof is complete.$\square$

Farkas' Lemma

An extremely important application of Theorem B.2 is to the issue of determining when a system of linear equations

$$AX = B$$

has a nonnegative solution $X \geq 0$, which is a problem of obvious importance. Indeed, this system has a nonnegative solution $X \geq 0$ if and only if B is in the image $A(\mathbb{R}_+^m)$ of the nonnegative orthant $\mathbb{R}_+^m$. Furthermore, since $A(\mathbb{R}_+^m)$ is a closed, convex cone in $\mathbb{R}^n$, Theorem B.1 implies that the system has a nonnegative solution if and only if

$$\langle \alpha, A(\mathbb{R}_+^m) \rangle \geq 0 \quad \Rightarrow \quad \langle \alpha, B \rangle \geq 0$$

for all $\alpha \in \mathbb{R}^n$. Now, if the columns of A are $A_1, \ldots, A_m$, then $\langle \alpha, A(\mathbb{R}_+^m) \rangle \geq 0$ if and only if $\langle \alpha, A_i \rangle \geq 0$ for all i; that is, if and only if $\alpha^t A \geq 0$, where α^t is the transpose of α. (We are writing our vectors as columns, so α^t is a row vector.) This important result is called *Farkas' lemma*.

Theorem B.3 (Farkas' Lemma) *Let A be an $n \times m$ real matrix and let $B \in \mathbb{R}^n$. The system*

$$AX = B$$

has a nonnegative solution X if and only if for all $\alpha \in \mathbb{R}^n$,

$$\alpha^t A \geq 0 \quad \Rightarrow \quad \alpha^t B \geq 0 \qquad\qquad \square$$

A commonly seen equivalent formulation of Farkas' lemma is the following, called a **theorem of the alternative** since it says that exactly one of two statements must hold.

Theorem B.4 (Farkas' Lemma) *Let A be an $n \times m$ real matrix and let $B \in \mathbb{R}^n$. Exactly one of the following holds:*

a) *The system*

$$AX = B$$

has a nonnegative solution.

b) *There is an $\alpha \in \mathbb{R}^n$ for which $\alpha^t A \geq 0$ and $\alpha^t B < 0$.*

Proof. Note that the negation of the implication in the first version of Farkas' lemma is statement b) above and 2a) and 2b) are mutually exclusive. $\square$

In some cases, we wish to know if a *subspace* S of $\mathbb{R}^m$ contains a nonnegative solution to the system $AX = B$; that is, if the set

$$K_{A,B}(S) = \{X \in S \mid AX = B, X \geq 0\}$$

is nonempty. To see how Farkas' lemma applies, let $\mathcal{Z} = \{Z_1, \ldots, Z_k\}$ be a basis for the orthogonal complement $S^{\perp}$ of S. Then $X \in S$ if and only if $\langle X, Z_i \rangle = 0$ for all i; that is, if and only if

$$NX = 0$$

where

$$N = \begin{pmatrix} Z_1 \\ Z_2 \\ \vdots \\ Z_k \end{pmatrix}$$

is the $k \times m$ matrix whose rows are the vectors in $\mathcal{Z}$.

Thus, $X \in K_{A,B}(S)$ if and only if X is a nonnegative solution to the expanded system

$$\begin{pmatrix} A \\ N \end{pmatrix} X = \begin{pmatrix} B \\ 0 \end{pmatrix}$$

Farkas' lemma now gives the following.

Theorem B.5 *Let A be an $n \times m$ real matrix and let $B \in \mathbb{R}^n$. Let S be a subspace of $\mathbb{R}^m$, let $\mathcal{Z} = \{Z_1, \ldots, Z_k\}$ be a basis for the orthogonal complement $S^{\perp}$ and let*

$$N = \begin{pmatrix} Z_1 \\ Z_2 \\ \vdots \\ Z_k \end{pmatrix}$$

be the $k \times m$ matrix whose rows are the vectors in $\mathcal{Z}$. Then exactly one of the following holds:

1) $K_{A,B}(S)$ *is nonempty; that is, the system $AX = B$ has a nonnegative solution in S.*

2)   *There is an $\alpha = \begin{pmatrix} \alpha_1 \\ \alpha_2 \end{pmatrix} \in \mathbb{R}^{n+k}$ with $\alpha_1 \in \mathbb{R}^n$ and $\alpha_2 \in \mathbb{R}^k$ for which*

$$(\alpha_1^t \quad \alpha_2^t) \begin{pmatrix} A \\ N \end{pmatrix} \geq 0 \quad and \quad \alpha_1^t B < 0$$

that is,

$$\alpha_1^t A + \alpha_2^t N \geq 0 \quad and \quad \alpha_1^t B < 0 \qquad\qquad \square$$

An Optimization Problem

Let A be an $n \times m$ real matrix and let S be a subspace of $\mathbb{R}^m$. As before, let $\mathcal{Z} = \{Z_1, \ldots, Z_k\}$ be a basis for the orthogonal complement $S^\perp$ and let

$$N = \begin{pmatrix} Z_1 \\ Z_2 \\ \vdots \\ Z_k \end{pmatrix}$$

be the $k \times m$ matrix whose rows are the vectors in $\mathcal{Z}$. We have seen that the set

$$\begin{pmatrix} A \\ N \end{pmatrix} (\mathbb{R}_+^m)$$

is a closed, convex cone in $\mathbb{R}^n$. Therefore, if $B_k \to B$ in $\mathbb{R}^n$ and if each of the systems

$$\begin{pmatrix} A \\ N \end{pmatrix} X = \begin{pmatrix} B_k \\ 0 \end{pmatrix} \qquad\qquad (B.2)$$

has a nonnegative solution X_k, it follows that the system

$$\begin{pmatrix} A \\ N \end{pmatrix} X = \begin{pmatrix} B \\ 0 \end{pmatrix}$$

also has a nonnegative solution. Now suppose that

$$A = \begin{pmatrix} A' \\ \alpha \end{pmatrix} \quad and \quad B_k = \begin{pmatrix} B' \\ b_k \end{pmatrix}$$

where α is the last row of A and b_k is the last coordinate of B_k. Then the system (B.2) can be written in the form

$$\begin{pmatrix} A' \\ N \end{pmatrix} X = \begin{pmatrix} B' \\ 0 \end{pmatrix}$$
$$\langle \alpha, X \rangle = b_k$$

In fact, if $f \colon \mathbb{R}^n \to \mathbb{R}$ is the linear functional whose Riesz vector is α, then (B.2) can be written in the form

$$\begin{pmatrix} A' \\ N \end{pmatrix} X = \begin{pmatrix} B' \\ 0 \end{pmatrix}$$
$$f(X) = b_k$$

Thus, if for each $k \geq 0$, there is a nonnegative solution X_k to this system and if $b_k \to b$, then there is also a nonnegative solution to the system

$$\begin{pmatrix} A' \\ N \end{pmatrix} X = \begin{pmatrix} B' \\ 0 \end{pmatrix}$$
$$f(X) = b$$

Now we can pose our optimization problem. Suppose that the set

$$K_{A',B'}(\mathcal{S}) = \{X \in \mathcal{S} \mid A'X = B', X \geq 0\}$$

is nonempty and that f is bounded from below on $K_{A',B'}(\mathcal{S})$; that is,

$$b = \min\{f(X) \mid X \in K_{A',B'}(\mathcal{S})\} > -\infty$$

Then there is a sequence $X_k \in K_{A',B'}(\mathcal{S})$ for which

$$b_k = f(X_k) \to b$$

Hence, the system

$$\begin{pmatrix} A' \\ N \end{pmatrix} X = \begin{pmatrix} B' \\ 0 \end{pmatrix}$$
$$f(X) = b$$

also has a nonnegative solution; that is, the system

$$A'X = B'$$
$$f(X) = b$$

has a solution in $K_{A',B'}(\mathcal{S})$. In other words, there is an element $X \in K_{A',B'}(\mathcal{S})$ at which f takes its *minimum value* on $K_{A',B'}(\mathcal{S})$. A similar statement can be made for the maximum value of f on $K_{A',B'}(\mathcal{S})$.

Theorem B.6 *Let A be an $n \times m$ real matrix, let $B \in \mathbb{R}^n$ and let $\mathcal{S}$ be a subspace of $\mathbb{R}^m$. Suppose that the set*

$$K_{A,B}(\mathcal{S}) = \{X \in \mathcal{S} \mid AX = B, X \geq 0\}$$

is nonempty. Let f be a linear functional on $\mathbb{R}^n$.
1)   If f is bounded from below on $K_{A,B}(\mathcal{S})$; that is, if

$$b = \min\{f(X) \mid X \in K_{A,B}(\mathcal{S})\} > -\infty$$

   then the system

$$AX = B$$
$$f(X) = b$$

has a nonnegative solution in S. In other words, there is an element $Y \in K_{A,B}(S)$ for which

$$f(Y) = \min\{f(X) \mid X \in K_{A,B}(S)\}$$
$$= \min\{f(X) \mid AX = B, X \in S, X \geq 0\}$$

2) If f is bounded from above on $K_{A,B}(S)$; that is, if

$$b = \max\{f(X) \mid X \in K_{A,B}(S)\} < \infty$$

then the system

$$AX = B$$
$$f(X) = b$$

has a nonnegative solution in S. In other words, there is an element $Y \in K_{A,B}(S)$ for which

$$f(Y) = \max\{f(X) \mid X \in K_{A,B}(S)\}$$
$$= \max\{f(X) \mid AX = B, X \in S, X \geq 0\}$$

In this context, the linear functional f to be maximized or minimized is called the **objective function** and the set $K_{A,B}(S)$ over which the objective function is to be optimized is called the **feasible region**. This theorem states that the optimal value (maximum or minimum) is achieved within the feasible region. $\square$

The Main Result

Now we are ready to address the main purpose of this appendix. We use the notation of Chapter 7. In particular, $\mathbb{M}$ is a discrete-time model, $\mathcal{M}$ is the subspace of all attainable vectors in $\mathbb{R}^m$, $\mathcal{I}$ is the initial pricing linear functional on $\mathcal{M}$, W is a nonattainable vector and

$$\mathcal{D}_W = \{X \in \mathcal{M} \mid X \geq W\}$$

is the set of attainable dominating vectors for W. Also, $\mathcal{E}_{\geq}(\mathcal{I})$ is the set of all nonnegative extensions of $\mathcal{I}$ to $\mathbb{R}^m$.

To deal with the fact that an element of $\mathcal{M}$ need not be nonnegative, we have the following lemma.

Lemma B.7 Let S be a subspace of $\mathbb{R}^n$ that contains a strongly positive vector. Then any vector in S can be written as the difference of two nonnegative vectors in S.

Proof. Let $X \in S$, let $B_1, \ldots, B_k$ be a basis for S consisting of strongly positive vectors and let $X = \Sigma x_i B_i$. Write

$$Y = \sum_{i:x_i>0} x_i B_i \quad \text{and} \quad Z = \sum_{i:x_i<0} x_i B_i$$

Then Y and $-Z$ are nonnegative in S and $X = Y - (-Z)$. $\square$

Thus, in view of Theorem 7.7, any vector in $\mathcal{M}$ can be written as the difference of two *nonnegative* vectors in $\mathcal{M}$.

The Minimum Dominating Price Is Attained

(*For convenience, we will write elements of* $\mathbb{R}^n$ *as either row vectors or column vectors.*) We wish to show using Theorem B.6 that there is a $Y \in \mathcal{D}_W$ for which

$$\mathcal{I}(Y) = \min\{\mathcal{I}(X) \mid X \in \mathcal{D}_W\}$$

The feasible set is

$$\mathcal{D}_W = \{X \in \mathcal{M} \mid X \geq W\}$$

but in order to apply Theorem B.6, we need a feasible set of the form

$$K_{A,B}(\mathcal{S}) = \{X \in \mathcal{S} \mid AX = B, X \geq 0\}$$

where A is a matrix. But in view of Lemma B.7, we can write each $X \in \mathcal{D}_W$ in the form $X = U - V$ for nonnegative $U, V \in \mathcal{M}$. Moreover, the fact that $X \geq W$ can be expressed by saying that there exists a nonnegative Z for which

$$W = X - Z = U - V - Z$$

This leads us to consider the feasible set

$$K_{A,W} = \{(U, V, Z) \in \mathcal{M} \times \mathcal{M} \times \mathbb{R}^m \mid U - V - Z = W, (U, V, Z) \geq 0\}$$

which does have the form $K_{A,B}(\mathcal{S})$ for

$$A = (I_m \; -I_m \; -I_m) \quad \text{and} \quad B = W$$

Note that for each $(U, V, Z) \in K_{A,W}$, we have $X = U - V \in \mathcal{D}_W$ and for each $X \in \mathcal{D}_W$, there is at least one triple $(U, V, Z) \in K_{A,W}$ for which $X = U - V$. Put another way, the function $\phi: K_{A,W} \to \mathcal{D}_W$ defined by

$$\phi(U, V, Z) = U - V$$

is *surjective*. In particular, $K_{A,W}$ is nonempty.

As to linear functionals, the objective function $\mathcal{I}$ is defined on $\mathcal{D}_W$, but we can define a corresponding linear functional $\mathcal{I}'$ on $K_{A,W}$ by

$$\mathcal{I}'(U, V, Z) = \mathcal{I}(U - V)$$

for $(U, V, Z) \in K_{A,W}$. It follows that the image of $\mathcal{I}'$ is the same as the image of $\mathcal{I}$, in symbols,

$$\{\mathcal{I}(X) \mid X \in \mathcal{D}_W\} = \{\mathcal{I}'(U, V, Z) \mid (U, V, Z) \in K_{A,W}\}$$

Hence, (B.1) implies that

$$-\infty < \min\{\mathcal{I}(X) \mid X \in D_W\} = \min\{\mathcal{I}'(U, V, Z) \mid (U, V, Z) \in K_{A,W}\}$$

But Theorem B.6 implies that the latter is taken on by an element (U', V', Z'); that is,

$$\mathcal{I}'(U', V', Z') = \min\{\mathcal{I}'(U, V, Z) \mid (U, V, Z) \in K_{A,W}\}$$

and so if $Y = U' - V'$, then

$$\mathcal{I}(Y) = \min\{\mathcal{I}(X) \mid X \in D_W\}$$

Theorem B.8 *Let $W \in \mathbb{R}^m$ be an alternative. Then there is a $Y \in \mathcal{D}_W$ for which*

$$\mathcal{I}(Y) = \min\{\mathcal{I}(X) \mid X \in D_W\} \qquad \qquad \square$$

The Maximum Extension Price Is Attained

The reader may wish to reread the discussion of the Riesz Representation Theorem (Theorem 7.3) before proceeding.

We wish to show next that there is a $g \in \mathcal{E}_{\gg}(\mathcal{I})$ for which

$$g(W) = \max\{f(W) \mid f \in \mathcal{E}_{\geq}(\mathcal{I})\}$$

On the surface, this does not look like it will fit the pattern of Theorem B.6, since we are maximizing over a collection of linear functionals. However, the Riesz bijection

$$\mathcal{R}(f) = Y_f$$

of Theorem 7.3 shows that we can work with the vectors in $\mathbb{R}^m$, thinking of them as Riesz vectors. We want to translate the problem at hand into the language of Riesz vectors. Let $\{A_1, \ldots, A_k\}$ be a basis for $\mathcal{M}$.

Now, a linear functional f is an extension of $\mathcal{I}$ if and only if

$$f(A_i) = \mathcal{I}(A_i)$$

for all $i = 1, \ldots, k$ or in terms of Riesz vectors,

$$\langle A_i, Y_f \rangle = \mathcal{I}(A_i)$$

for all $i = 1, \ldots, k$. Put another way, if

$$A = \begin{pmatrix} A_1 \\ A_2 \\ \vdots \\ A_k \end{pmatrix} \quad \text{and} \quad B = \begin{pmatrix} \mathcal{I}(A_1) \\ \mathcal{I}(A_2) \\ \vdots \\ \mathcal{I}(A_k) \end{pmatrix}$$

then this is equivalent to the system

$$AY_f = B$$

Furthermore, $f \geq 0$ if and only if $Y_f \geq 0$. Hence, the image of the set $\mathcal{E}_{\geq}(\mathcal{I})$ under the Riesz map is the set

$$K_{A,B} = \{Y_f \in \mathbb{R}^m \mid AY_f = B, X \geq 0\}$$

where Y_f represents any vector in $\mathbb{R}^m$. In other words, the correspondence

$$f \leftrightarrow Y_f$$

is a *bijection* between $\mathcal{E}_{\geq}(\mathcal{I})$ and $K_{A,B}$. In particular, $K_{A,B}$ is nonempty.

Now, we want to maximize $f(W)$ over $f \in \mathcal{E}_{\geq}(\mathcal{I})$. To rephrase this in terms of Riesz vectors, define the map $\epsilon_W \colon \mathbb{R}^m \to \mathbb{R}$ by

$$\epsilon_W(Y_f) = f(W)$$

for all Riesz vectors $Y_f \in \mathbb{R}^m$ (that is, for all vectors in $\mathbb{R}^m$). Then maximizing $f(W)$ over all $f \in \mathcal{E}_{\geq}(\mathcal{I})$ is equivalent to maximizing $\epsilon_W(Y_f)$ over all $Y_f \in K_{A,B}$; that is,

$$\infty > \max\{f(W) \mid f \in \mathcal{E}_{\geq}(\mathcal{I})\} = \max\{\epsilon_W(Y_f) \mid Y_f \in K_{A,B}\}$$

Hence, Theorem B.6 implies that there is a $Y_g \in \mathbb{R}^m$ for which

$$\epsilon_W(Y_g) = \max\{\epsilon_W(Y_f) \mid Y_f \in K_{A,B}\}$$

and so

$$g(W) = \max\{f(W) \mid f \in \mathcal{E}_{\geq}(\mathcal{I})\}$$

Theorem B.9 *Let $W \in \mathbb{R}^m$ be an alternative. Then there is a $g \in \mathcal{E}_{\geq}(\mathcal{I})$ for which*

$$g(W) = \max\{f(W) \mid f \in \mathcal{E}_{\geq}(\mathcal{I})\} \qquad \square$$

The Minimum Dominating Price Equals the Maximum Extension Price

We can now show that the minimum dominating price, which can be expressed as

$$\mu_W = \min\{\mathcal{I}(U - V) \mid (U, V, Z) \in K_{A,W}\}$$

with

$$K_{A,W} = \{(U, V, Z) \in \mathcal{M} \times \mathcal{M} \times \mathbb{R}^m \mid A(U\ V\ Z)^t = W, (U, V, Z) \geq 0\}$$

where $A = (I_m\ -I_m\ -I_m)$ is equal to the maximum extension price

$$\sigma_W = \max\{f(W) \mid f \in \mathcal{E}_{\geq}(\mathcal{I})\}$$

We have already seen that

$$\sigma_W \leq \mu_W$$

Let N be the matrix for which

$$\mathcal{M} = \{X \in \mathbb{R}^m \mid NX = 0\}$$

If we show that there is an element (U, V, Z) in $K_{A,W}$ for which

$$\mathcal{I}(U - V) = \sigma_W$$

then

$$\sigma_W \leq \mu_W \leq \mathcal{I}(U - V) = \sigma_W$$

and so $\sigma_W = \mu_W$, as desired. To this end, let A' be the matrix defined by

$$A' = \begin{pmatrix} I_m & -I_m & -I_m \\ N & 0 & 0 \\ 0 & N & 0 \\ Y_{\mathcal{I}} & -Y_{\mathcal{I}} & 0 \end{pmatrix}$$

where $Y_{\mathcal{I}}$ is the Riesz vector for $\mathcal{I}$. This matrix has $m + 2k + 1$ rows and $3m$ columns, where $k = \dim(\mathcal{M}^\perp)$. We also let

$$B = \begin{pmatrix} W \\ 0 \\ 0 \\ \sigma_W \end{pmatrix}$$

Then

$$A' \begin{pmatrix} U \\ V \\ Z \end{pmatrix} = \begin{pmatrix} U - V - Z \\ NU \\ NV \\ \mathcal{I}(U - V) \end{pmatrix}$$

We want to know if there is a nonnegative solution to the system

$$A' \begin{pmatrix} U \\ V \\ Z \end{pmatrix} = B$$

that is,

$$\begin{pmatrix} U - V - Z \\ NU \\ NV \\ \mathcal{I}(U - V) \end{pmatrix} = \begin{pmatrix} W \\ 0 \\ 0 \\ \sigma_W \end{pmatrix}$$

since then $(U, V, Z) \in K_{A,W}$ and $\mathcal{I}(U - V) = \sigma_W$, as desired.

To this end, we use Theorem B.5. Suppose there is an

$$\alpha = (\alpha, \alpha_1, \alpha_2, \beta) \in \mathbb{R}^m \times \mathbb{R}^k \times \mathbb{R}^k \times \mathbb{R}$$

for which

$$(\alpha^t, \alpha_1^t, \alpha_2^t, \beta) A' \geq 0 \tag{B.3}$$

We must show that $(\alpha^t, \alpha_1^t, \alpha_2^t, \beta) B \geq 0$; that is,

$$\alpha^t W + \beta \sigma_W \geq 0 \tag{B.4}$$

If $\Delta = (U, V, Z) \in \mathbb{R}^m \times \mathbb{R}^m \times \mathbb{R}^m$ is nonnegative, then (B.3) implies that

$$(\alpha^t, \alpha_1^t, \alpha_2^t, \beta) A' \begin{pmatrix} U \\ V \\ Z \end{pmatrix} \geq 0$$

that is,

$$(\alpha^t, \alpha_1^t, \alpha_2^t, \beta) \begin{pmatrix} U - V - Z \\ NU \\ NV \\ \mathcal{I}(U - V) \end{pmatrix} \geq 0$$

Now, we take some special cases. First, for $\Delta = (0, 0, Z)$ where $Z \geq 0$, we get

$$-\alpha^t Z \geq 0$$

for all $Z \geq 0$. Second, for any $X \in \mathcal{M}$ that has the form $X = U - V$ with $U, V \in \mathcal{M}$ nonnegative, we get for $\Delta = (U, V, 0)$,

$$\alpha^t X + \beta \mathcal{I}(X) \geq 0$$

Since this holds for all $X \in \mathcal{M}$, it holds for all $-X \in \mathcal{M}$ and so

$$\alpha^t X + \beta \mathcal{I}(X) = 0$$

Finally, any $X \in \mathcal{D}_W$ has the form $X = U - V$ where $U, V \in M$ are nonnegative and so for $\Delta = (U, V, Z) \in K_{A,W}$, we get

$$\alpha^t W + \beta \mathcal{I}(X) \geq 0$$

Let us summarize:

1) $-\alpha^t Z \geq 0$ for all $Z \geq 0$
2) $\alpha^t X + \beta \mathcal{I}(X) = 0$ for all $X \in M$
3) $\alpha^t W + \beta \mathcal{I}(X) \geq 0$ for all $X \in \mathcal{D}_W$.

Now, if $\beta \leq 0$, then for all $X \in \mathcal{D}_W$,

$$\sigma_W \leq \mu_W \leq \mathcal{I}(X)$$

implies that $\beta \mathcal{I}(X) \leq \beta \sigma_W$ and so 3) gives

$$0 \leq \alpha^t W + \beta \mathcal{I}(X) \leq \alpha^t W + \beta \sigma_W$$

which gives (B.4), as desired.

On the other hand, if $\beta > 0$, then 2) gives

$$\mathcal{I}(X) = -\frac{1}{\beta} \langle \alpha, X \rangle$$

and so the linear functional $f_\alpha : \mathbb{R}^m \to \mathbb{R}$ defined by

$$f_\alpha(X) = -\frac{1}{\beta} \langle \alpha, X \rangle$$

is an extension of $\mathcal{I}$. Moreover, 1) implies that f is nonnegative and so $f \in \mathcal{E}_\geq(\mathcal{I})$. It follows that

$$\sigma_W \geq f(W) = -\frac{1}{\beta} \langle \alpha, W \rangle = -\frac{1}{\beta} \alpha^t W$$

which is also (B.4). This completes the proof that $\mu_W = \sigma_W$.

We can now summarize.

Theorem B.10 *Let $W \in \mathbb{R}^m$ be an alternative.*
1) *There is a $Y \in \mathcal{D}_W$ for which*

$$\mathcal{I}(Y) = \mu_W = \min\{\mathcal{I}(X) \mid X \in D_W\}$$

2) *There is a $g \in \mathcal{E}_\geq(\mathcal{I})$ for which*

$$g(W) = \sigma_W = \max\{f(W) \mid f \in \mathcal{E}_\geq(\mathcal{I})\}$$

3) *The minimum dominating price of W is equal to the maximum extension price of W; that is,*

$$\mu_W = \sigma_W \qquad \qquad \square$$

Selected Solutions

Chapter 1

3. $2C_{100} - 2C_{105} - 0.5C_{150} + 0.5C_{170}$
4. The cost C_1 of the call with the smaller strike price is more than the cost C_2 of the call with the higher strike price. The profit curve is shown below.

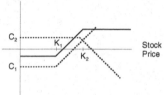

Chapter 2

1. We still have

$$V(\text{long contract}) = S_T - F_{0,T}$$
$$V(\text{short contract}) = F_{0,T} - S_T$$

but the cash-and-carry investor has a final payoff of

$$V(\text{cash-and-carry}) = S_T - S_0 e^{rT} + I e^{rT}$$

and

$$V(\text{reverse cash-and-carry}) = S_0 e^{rT} - S_T - I e^{rT}$$

To explain the last term, note that the short sale of an asset requires a lender to lend that asset. This lender will demand the return of not only the asset itself, but also the income that would have come to the lender by virtue of owning the asset. The two strategies now have payoffs as follows:

$$V(\text{long contract}) + V(\text{reverse cash-and-carry}) = S_0 e^{rT} - F_{0,T} - I e^{rT}$$

and

$$\mathcal{V}(\text{short contract}) + \mathcal{V}(\text{cash-and-carry}) = F_{0,T} - S_0 e^{rT} + I e^{rT}$$

Setting the final payoffs to 0 gives

$$F_{0,T} = (S_0 - I) e^{rT}$$

3. If not then buy the share, sell the call and pocket the difference. Use the share to cover the call if and when it is exercised.

7. Consider two portfolios. Portfolio A is one put. Portfolio B is $Ke^{-rT} + d_0$ in cash and a short share of stock. Then

$$\mathcal{V}_{A,0} = P \quad \text{and} \quad \mathcal{V}_{B,0} = Ke^{-rT} + d_0 - S_0$$

and

$$\mathcal{V}_{A,T} = (K - S_T)^+ \quad \text{and} \quad \mathcal{V}_{B,T} = K + d_0 e^{rT} - S_T - d_0 e^{rT} = K - S_T$$

Since $\mathcal{V}_{A,T} \geq \mathcal{V}_{B,T}$, it follows that $\mathcal{V}_{A,0} \geq \mathcal{V}_{B,0}$.

Chapter 3

1. $1/2$
3. $5/12, 9/12, 7/12$
5. $1/3, 1/6, 1/3, 1/2$
7. $11/16$
9. $0.28; 0.55; 0.6$
13. $1/2$; yes.
15. $25/13$ cents. Yes.

Chapter 4

5. The linearity of conditional expectation implies that for any $B_k \in \mathcal{P}_k$,

$$\mathcal{E}(X_{k+1} + Y_{k+1} \mid B_k) = \mathcal{E}(X_{k+1} \mid B_k) + \mathcal{E}(Y_{k+1} \mid B_k)$$
$$= X_k(B_k) + Y_k(B_k)$$
$$= 0$$

and so the local martingale condition holds for $\mathbb{X} + \mathbb{Y}$.

7. For part a), since $\mathbb{M}$ is a martingale, we have

$$\mathcal{E}(M_{k+1} \mid \mathcal{P}_k) - \mathcal{E}(M_k \mid \mathcal{P}_k) = \mathcal{E}(M_{k+1} \mid \mathcal{P}_k) - M_k = 0$$

and since $\mathbb{A}$ is predictable,

$$\mathcal{E}(A_{k+1} \mid \mathcal{P}_k) - \mathcal{E}(A_k \mid \mathcal{P}_k) = A_{k+1} - A_k$$

and so taking conditional expectation gives

$$\mathcal{E}(X_{k+1} \mid \mathcal{P}_k) - X_k = A_{k+1} - A_k$$

Now we sum from $k = 0$ to $k = n - 1$ to get

$$\sum_{k=0}^{n-1}(A_{k+1} - A_k) = \sum_{k=0}^{n-1}(\mathcal{E}(X_{k+1} \mid \mathcal{P}_k) - X_k)$$

The left side is telescoping and since $A_0 = 0$, we get

$$A_n = \sum_{k=0}^{n-1}(\mathcal{E}(X_{k+1} \mid \mathcal{P}_k) - X_k)$$

Then we can set $M_n = X_n - A_n$. It follows that

$$\begin{aligned}
\mathcal{E}(A_n \mid \mathcal{P}_{n-1}) &= \sum_{k=0}^{n-1}\mathcal{E}(\mathcal{E}(X_{k+1} \mid \mathcal{P}_k) - X_k \mid \mathcal{P}_{n-1}) \\
&= \sum_{k=0}^{n-1}(\mathcal{E}(X_{k+1} \mid \mathcal{P}_k) - \mathcal{E}(X_k \mid \mathcal{P}_{n-1})) \\
&= \sum_{k=0}^{n-1}(\mathcal{E}(X_{k+1} \mid \mathcal{P}_k) - X_k) \\
&= A_n
\end{aligned}$$

and so $\mathbb{A}$ is indeed predictable. Moreover,

$$\begin{aligned}
\mathcal{E}(M_n \mid \mathcal{P}_{n-1}) &= \mathcal{E}(X_n \mid \mathcal{P}_{n-1}) - A_n \\
&= \mathcal{E}(X_n \mid \mathcal{P}_{n-1}) - \sum_{k=0}^{n-1}(\mathcal{E}(X_{k+1} \mid \mathcal{P}_k) - X_k) \\
&= X_{n-1} - \sum_{k=0}^{n-2}(\mathcal{E}(X_{k+1} \mid \mathcal{P}_k) - X_k) \\
&= X_{n-1} - A_{n-1} \\
&= M_{n-1}
\end{aligned}$$

and so $\mathbb{M}$ is a martingale.

For part b), suppose that $\mathbb{X}$ is a supermartingale. Then

$$\begin{aligned}
M_k - A_k = X_k &\geq \mathcal{E}(X_{k+1} \mid \mathcal{P}_k) \\
&= \mathcal{E}(M_{k+1} \mid \mathcal{P}_k) - \mathcal{E}(A_{k+1} \mid \mathcal{P}_k) \\
&= M_k - A_{k+1}
\end{aligned}$$

and so $A_k \leq A_{k+1}$. The case for submartingales is proved similarly.

Chapter 5

1. Consider a model with just one time interval $[t_0, t_1]$ and two risky assets $\mathfrak{a}_2$ and $\mathfrak{a}_3$. Assume that $\Omega = \{\omega_1, \omega_2\}$ and let the prices be

$$S_{0,2} = 1, \quad S_{1,2}(\omega) = \begin{cases} 2 & \text{for } \omega = \omega_1 \\ 0 & \text{for } \omega = \omega_2 \end{cases}$$

and

$$S_{0,3} = 1, \quad S_{1,3}(\omega) = \begin{cases} 0 & \text{for } \omega = \omega_1 \\ 3 & \text{for } \omega = \omega_2 \end{cases}$$

Then neither asset by itself can form an arbitrage opportunity but the trading strategy Φ defined by

$$\Theta_1 = (0, 1.5, 1)$$

has values (we assume that the risk-free rate is 0)

$$\overline{V}_0(\Phi) = 2.5 \quad \text{and} \quad \overline{V}_1(\Phi) = \begin{cases} 3 & \text{for } \omega = \omega_1 \\ 3 & \text{for } \omega = \omega_2 \end{cases}$$

and so Φ is an arbitrage trading strategy.

3.   The system of equations is

$$\mathcal{V}_3(\Theta_3)(\omega_1) = 100$$
$$\mathcal{V}_3(\Theta_3)(\omega_2) = 100$$
$$\mathcal{V}_3(\Theta_3)(\omega_3) = 95$$
$$\mathcal{V}_3(\Theta_3)(\omega_4) = 90$$
$$\mathcal{V}_3(\Theta_3)(\omega_5) = 90$$
$$\mathcal{V}_3(\Theta_3)(\omega_6) = 85$$

or, since the price of the risk-free asset a_1 is 1

$$\theta_{3,1}(\omega_1) + S_{3,2}(\omega_1)\theta_{3,2}(\omega_1) = 100$$
$$\theta_{3,1}(\omega_2) + S_{3,2}(\omega_2)\theta_{3,2}(\omega_2) = 100$$
$$\theta_{3,1}(\omega_3) + S_{3,2}(\omega_3)\theta_{3,2}(\omega_3) = 95$$
$$\theta_{3,1}(\omega_4) + S_{3,2}(\omega_4)\theta_{3,2}(\omega_4) = 90$$
$$\theta_{3,1}(\omega_5) + S_{3,2}(\omega_5)\theta_{3,2}(\omega_5) = 90$$
$$\theta_{3,1}(\omega_6) + S_{3,2}(\omega_6)\theta_{3,2}(\omega_6) = 85$$

Substituting the actual prices gives

$$\theta_{3,1}(\omega_1) + 100\theta_{3,2}(\omega_1) = 100$$
$$\theta_{3,1}(\omega_2) + 95\theta_{3,2}(\omega_2) = 100$$
$$\theta_{3,1}(\omega_3) + 95\theta_{3,2}(\omega_3) = 95$$
$$\theta_{3,1}(\omega_4) + 90\theta_{3,2}(\omega_4) = 90$$
$$\theta_{3,1}(\omega_5) + 90\theta_{3,2}(\omega_5) = 90$$
$$\theta_{3,1}(\omega_6) + 80\theta_{3,2}(\omega_6) = 85$$

The condition that Θ_3 be $\mathcal{P}_2$-measurable is

$$\theta_{3,1}(\omega_1) = \theta_{3,1}(\omega_2)$$
$$\theta_{3,1}(\omega_4) = \theta_{3,1}(\omega_5)$$
$$\theta_{3,2}(\omega_1) = \theta_{3,2}(\omega_2)$$
$$\theta_{3,2}(\omega_4) = \theta_{3,2}(\omega_5)$$

and so the previous system can be written using only $\omega_1, \omega_3, \omega_4$ and ω_6 as

$$\theta_{3,1}(\omega_1) + 100\theta_{3,2}(\omega_1) = 100$$
$$\theta_{3,1}(\omega_1) + 95\theta_{3,2}(\omega_1) = 100$$
$$\theta_{3,1}(\omega_3) + 95\theta_{3,2}(\omega_3) = 95$$
$$\theta_{3,1}(\omega_4) + 90\theta_{3,2}(\omega_4) = 90$$
$$\theta_{3,1}(\omega_4) + 90\theta_{3,2}(\omega_4) = 90$$
$$\theta_{3,1}(\omega_6) + 80\theta_{3,2}(\omega_6) = 85$$

The first two equations have a unique solution and so do the fourth and fifth equations, giving

$$\Theta_3(\omega_1) = \Theta_3(\omega_2) = (100, 0)$$
$$\Theta_3(\omega_4) = \Theta_3(\omega_5) = (0, 1)$$

along with

$$\Theta_3(\omega_3) = \left(s, \frac{95 - s}{95} \right)$$
$$\Theta_3(\omega_6) = \left(t, \frac{85 - t}{80} \right)$$

where s and t are parameters. The acquisition values for Θ_3 are

$$\mathcal{V}_2(\Theta_3)(\omega_1) = \mathcal{V}_2(\Theta_3)(\omega_2) = 100$$
$$\mathcal{V}_2(\Theta_3)(\omega_3) = s + 80 \cdot \frac{95 - s}{95} = \frac{3s}{19} + 80$$
$$\mathcal{V}_2(\Theta_3)(\omega_4) = \mathcal{V}_2(\Theta_3)(\omega_5) = 80$$
$$\mathcal{V}_2(\Theta_3)(\omega_6) = t + 75 \cdot \frac{95 - t}{95} = \frac{4t}{19} + 75$$

The self-financing condition requires that these are also the liquidation values of Θ_2 and so Θ_2 must replicate the alternative

$$\left(100, \frac{3s}{19} + 80, 80, \frac{4t}{19} + 75 \right)$$

Since we are asked for only one replicating portfolio, let us choose $s = t = 0$ to get the alternative

$$(100, 80, 80, 75)$$

We have the system

$$\theta_{2,1}(\omega_1) + 90\theta_{2,2}(\omega_1) = 100$$
$$\theta_{2,1}(\omega_2) + 90\theta_{2,2}(\omega_2) = 100$$
$$\theta_{2,1}(\omega_3) + 80\theta_{2,2}(\omega_3) = 80$$
$$\theta_{2,1}(\omega_4) + 80\theta_{2,2}(\omega_4) = 80$$
$$\theta_{2,1}(\omega_5) + 80\theta_{2,2}(\omega_5) = 80$$
$$\theta_{2,1}(\omega_6) + 75\theta_{2,2}(\omega_6) = 75$$

Since $\theta_{2,i}$ is constant on the blocks $\{\omega_1, \omega_2, \omega_3\}$ and $\{\omega_6, \omega_6, \omega_6\}$ this can be written

$$\theta_{2,1}(\omega_1) + 90\theta_{2,2}(\omega_1) = 100$$
$$\theta_{2,1}(\omega_1) + 90\theta_{2,2}(\omega_1) = 100$$
$$\theta_{2,1}(\omega_1) + 80\theta_{2,2}(\omega_1) = 80$$
$$\theta_{2,1}(\omega_4) + 80\theta_{2,2}(\omega_4) = 80$$
$$\theta_{2,1}(\omega_4) + 80\theta_{2,2}(\omega_4) = 80$$
$$\theta_{2,1}(\omega_4) + 75\theta_{2,2}(\omega_4) = 75$$

or

$$\theta_{2,1}(\omega_1) + 90\theta_{2,2}(\omega_1) = 100$$
$$\theta_{2,1}(\omega_1) + 80\theta_{2,2}(\omega_1) = 80$$
$$\theta_{2,1}(\omega_4) + 80\theta_{2,2}(\omega_4) = 80$$
$$\theta_{2,1}(\omega_4) + 75\theta_{2,2}(\omega_4) = 75$$

giving

$$\Theta_2(\omega_1) = \Theta_2(\omega_2) = \Theta_2(\omega_3) = (-80, 2)$$
$$\Theta_2(\omega_4) = \Theta_2(\omega_5) = \Theta_2(\omega_6) = (0, 1)$$

Working backward in time, we next compute the acquisition values for Θ_2

$$\mathcal{V}_1(\Theta_2)(\omega_1) = -80 + 85 \cdot 2 = 90$$
$$\mathcal{V}_1(\Theta_2)(\omega_4) = 0 + 78 = 78$$

The self-financing condition requires that these are also the liquidation values of Θ_1 and so

$$\mathcal{V}_1(\Theta_1)(\omega_1) = 90$$
$$\mathcal{V}_1(\Theta_1)(\omega_4) = 78$$

Writing these out and substituting the actual prices gives the system

$$\theta_{1,1}(\omega_1) + 85\theta_{1,2}(\omega_1) = 90$$
$$\theta_{1,1}(\omega_4) + 78\theta_{1,2}(\omega_4) = 78$$

But Θ_1 is $\mathcal{P}_0$-measurable; that is, constant on Ω, and so for any $\omega \in \Omega$

$$\theta_{1,1}(\omega_1) + 85\theta_{1,2}(\omega_1) = 90$$
$$\theta_{1,1}(\omega_4) + 78\theta_{1,2}(\omega_4) = 78$$

This system has solution

$$\Theta_1(\omega) = \left(-\frac{390}{7}, \frac{12}{7}\right)$$

which is a portfolio consisting of a short position (sale) of $\frac{390}{7}$ bonds and a purchase of $\frac{12}{7}$ shares of stock, for an initial cost of

$$-\frac{390}{7} + 80 \cdot \frac{12}{7} = \frac{570}{7} \approx \$81.43$$

5. Toss 1 is tails: Casino is even, player down \$1 million, game over.
    Toss 1 is heads: Casino up \$2 million, player down \$1 million, game continues.
    Toss 2 is tails: Casino is even, player down \$1 million, game over.
    Toss 2 is heads: Casino up \$4 million, player down \$1 million, game continues.
    Toss 3 is tails: Casino is even, player down \$1 million, game over.
    Toss 3 is heads: Casino is even, player up \$8 million, game continues.

    In all ending cases, the casino is even. Thus, the casino has a perfect hedge. It is self-financing because the casino never added its own money or removed money. The side bets on heads replicated the payoff to the casino, but in the opposite position, resulting in a 0 payoff to the casino.

7. The self-financing condition is

$$V_i(\Theta'_i) = V_i(\Theta'_{i+1})$$

for all $i = 1, \ldots, T-1$. Because Φ is assumed to be self-financing, the liquidation value of Θ'_i is

$$V_i(\Theta'_i) = (\theta_{i,1} + a)S_{i,1} + \sum_{j=2}^{n}\theta_{i,j}S_{i,j}$$
$$= V_i(\Theta_i) + aS_{i,1}1_\Omega$$
$$= V_i(\Theta_{i+1}) + aS_{i,1}1_\Omega$$

and the acquisition value is

$$V_i(\Theta'_{i+1}) = (\theta_{i+1,1} + a)S_{i,1} + \sum_{j=2}^{n}\theta_{i+1,j}S_{i,j}$$
$$= V_i(\Theta_{i+1}) + aS_{i,1}1_\Omega$$

Thus Φ' is self-financing.

Chapter 6

1. a) 0, b) 0.0015, c) 0.2944, d) .0816, e) 2.0799, f) .0783. For the put, use the put-call option parity formula $P = Ke^{-rt} + C - S_0$. For example, when $K = 50$ we have

$$P = 50e^{0.01/6} + 0.2944 - 50 = 0.3778$$

3. A 10% gain followed by a 10% loss, or vice versa, results is a slight loss, as shown by

$$(1 + 0.1)(1 - 0.1) = 1 - 0.01 = 0.99$$

(If the gain comes first, the loss is on a larger amount; if the loss comes first, the gain is on a smaller amount.

Chapter 7

1. Let $f : \mathbb{R}^2 \to \mathbb{R}$ be defined by $f(E_1) = 1, f(E_2) = 0$.
5. A vector $(2a, 3a) \in \mathcal{M}$ dominates (x, y) if and only if $x \leq 2a$ and $y \leq 3a$; that is, if and only if $a \geq \max\{x/2, y/3\}$. Thus, the minimum dominating price is achieved by $\mathcal{I}(2a, 3a) = a\alpha$ for the smallest possible a, which is $a = \max\{x/2, y/3\}$.

Chapter 8

3. We have

$$[\tau = k] = \{\omega \mid S_k \in B \text{ but } S_j \in B^c \text{ for } j < k\}$$
$$= [S_0 \in B^c] \cup \cdots \cup [S_{k-1} \in B^c] \cup [S_k \in B]$$

But since the price S_i is $\mathcal{P}_i$-measurable and since $(\mathcal{P}_i)$ is a filtration, we deduce that each of the events $[S_i \in B^c]$ and the event $[S_k \in B]$ are in the largest algebra $\mathcal{A}(\mathcal{P}_k)$. This is the condition required of a stopping time. Finally, for $k = T$ we have

$$[\tau = T] = [S_0 \in B^c] \cap \cdots \cap [S_{T-1} \in B^c] \in \mathcal{A}(\mathcal{P}_T)$$

5. Write

$$\{k \mid S_k(\omega) \geq 2S_{k-1}(\omega)\} = \left\{k \mid \frac{S_k(\omega)}{S_{k-1}(\omega)} \geq 2\right\}$$

and consider first entry times for the adapted process $\left(\frac{S_k(\omega)}{S_{k-1}(\omega)}\right)$.

Chapter 9

5. If $a \in \mathbb{R}$ then

$$(-\infty, a) = \bigcup_{n > a} (-n, a)$$

is a countable union of open intervals and so is a Borel set. Also,

$$(-\infty, a] = \bigcap_{n>0} \left(-\infty, a + \frac{1}{n}\right) \in \mathcal{B}$$

The right rays are complements of the left rays.

13. Suppose that $A_1 \subseteq A_2 \subseteq \cdots$ is an increasing sequence of events and let

$$A = \bigcup_{i=1}^{\infty} A_i$$

The limit exists because it is the limit of an increasing bounded sequence of real numbers. Set $A_0 = \emptyset$ and write

$$A = \bigcup_{i=1}^{\infty} (A_i \setminus A_{i-1})$$

where the events $A_i \setminus A_{i-1}$ are disjoint. Then

$$\mathbb{P}(A) = \mathbb{P}\left(\bigcup_{i=1}^{\infty} (A_i \setminus A_{i-1})\right)$$

$$= \sum_{i=1}^{\infty} \mathbb{P}(A_i \setminus A_{i-1})$$

$$= \lim_{n\to\infty} \sum_{i=1}^{n} \mathbb{P}(A_i \setminus A_{i-1})$$

$$= \lim_{n\to\infty} [\mathbb{P}(A_i) - \mathbb{P}(A_0)]$$

$$= \lim_{n\to\infty} \mathbb{P}(A_i)$$

15. For part b)

$$\mathbb{P}((a, b]) = \mathbb{P}\left(\bigcup_{n=1}^{\infty} (a, b - \frac{1}{n}]\right)$$

$$= \lim_{n\to\infty} \mathbb{P}\left((a, b - \frac{1}{n}]\right)$$

$$= \lim_{n\to\infty} F\left(b - \frac{1}{n}\right) - F(a)$$

$$= F(b-) - F(a)$$

Chapter 10

1. $\mu = 0.15, \sigma^2 = 0.03$.
3. $C = \$33.36$.

5. For the expected value, we have

$$\mathcal{E}_p(Q_s) = \mathcal{E}_p\left(\sigma_s\sqrt{\Delta t}\sum_{i=1}^{T}X_{s,i}\right)$$

$$= \sigma_s\sqrt{\Delta t}\sum_{i=1}^{T}\mathcal{E}_p(X_{s,i})$$

$$= \frac{\sigma_s\sqrt{\Delta t}}{\sqrt{s(1-s)}}\sum_{i=1}^{T}[(1-s)p - s(1-p)]$$

$$= \frac{\sigma_s\sqrt{\Delta t}}{\sqrt{s(1-s)}}T(p-s)$$

$$= \sigma_s\frac{t}{\sqrt{\Delta t}}\frac{p-s}{\sqrt{s(1-s)}}$$

References

The Mathematics of Finance

1. Baxter, M. and Rennie, A., *Financial Calculus: An Introduction to Derivative Pricing*, Cambridge University Press, 1996, 0-521-55289-3.
2. Etheridge, A., *A Course in Financial Calculus*, Cambridge University Press, 2002, 0-521-89077-2.
3. Föllmer, H., *Stochastic Finance: An Introduction in Discrete Time*, Walter de Gruyter, 2002, 3-11-017119-8.
4. Kallianpur, G. and Karandikar, R., *Introduction to Option Pricing Theory*, Birkhauser, 1999, 0-8176-4108-4.
5. Lo, A. and MacKinlay, C., *A Non-Random Walk Down Wall Street*, Princeton University Press, 2001, 0-691-09256-7.
6. Medina, P. and Merino, S., *Mathematical Finance and Probability*, Birkhauser, 2003, 3-7643-6921-3.
7. Ross, S., *An Introduction to Mathematical Finance: Options and Other Topics*, Cambridge University Press, 1999, 0-521-77043-2.
8. Ross, S., *An Elementary Introduction to Mathematical Finance*, Cambridge University Press, 2002, 0-521-81429-4.
9. Shafer, G, and Vovk, V., *Probability and Finance: It'gs Only a Game*, John Wiley, 0-471-40226-5.
10. Wilmott, P., Howison, S. and Dewynne, J., *The Mathematics of Financial Derivatives: A Student Introduction*, Cambridge University Press, 1995, 0-521-49789-2.

Probability

1. Chung, K., *A Course in Probability Theory*, Academic Press, 2001, 0-12-174151-6.
2. Grimmett, Geoffrey, *Probability*, Oxford University Press, 1986, 0-19-853264-4.
3. Grimmett, Geoffrey R., *One Thousand Exercises in Probability*, Oxford University Press, 2001, 0-19-857221-2.
4. Grimmett, G. and Stirzaker, D., *Probability and Random Processes*, Oxford University Press, 2001, 0-19-857222-0.

5. Lewis, H. W., *Why Flip a Coin?: The Art and Science of Good Decisions*, Wiley, 1997, 0-471-16597-2.
6. Karr, Alan, *Probability*, Springer, 1993, 0-387-94071-5.
7. Issac, Richard, *The Pleasures of Probability*, Springer, 1995, 0-387-94415-X.
8. Shiryaev, Albert N., *Probability*, Springer, 1995, 0-387-94549-0.
9. Gordon, Hugh, *Discrete Probability*, Springer, 1997, 0-387-98227-2.
10. Billingsley, Patrick, *Probability and Measure*, Wiley, 1995, 0-471-00710-2.
11. Feller, William, *An Introduction to Probability Theory and Its Applications*, 3E, Vol. 1, Wiley, 1968, 0-471-25708-7.
12. Feller, William, *An Introduction to Probability Theory and Its Applications*, 2E, Vol. 2, Wiley, 1971, 0-471-25709-5.
13. Mises, Richard von, *Probability, Statistics And Truth*, Dover, 0-486-24214-5.
14. Heathcote, C. R., *Probability: Elements of the Mathematical Theory*, Dover, 0-486-41149-4.
15. Goldberg, Samuel, *Probability: An Introduction*, Dover, 0-486-65252-1.
16. Mosteller, Frederick, *Fifty Challenging Problems in Probability with Solutions*, Dover, 0-486-65355-2.
17. Freund, John E., *Introduction To Probability*, Dover, 0-486-67549-1.
18. Brieman, L., *Probability*, SIAM, 1992, 0-89871-296-3.
19. Brzezniak, Z. and Zastawniak,T., *Basic Stochastic Processes*, Springer, 1998, 3-540-76175-6.

Options and Other Derivatives

1. Chriss, Neil, *Black–Scholes and Beyond*, McGraw-Hill, 1997, 0-7863-1025-1.
2. Hull, John, *Options, Futures and Other Derivatives*, Prentice Hall, Fifth Edition, 2003, 0-13-009056-5.
3. Irwin, R., *Option Volatility & Pricing*, McGraw-Hill, 1994, 1-55738-486-X.
4. McMillam, L., *Options as a Strategic Investment*, New York Institute of Finance, 2002, 0-7352-0197-8.
5. Thomsett, M., *Getting Started in Options*, Wiley, 2001, 0-471-40946-4.

Index